AI UNLEASHED

AI UNLEASHED

A LEADER'S PLAYBOOK TO MASTER AI FOR BUSINESS EXCELLENCE

REDDY MALLIDI

MANUSCRIPTS
PRESS

COPYRIGHT © 2024 REDDY MALLIDI
All rights reserved.

AI UNLEASHED

A Leader's Playbook to Master AI for Business Excellence

ISBN

979-8-88926-174-2   *Paperback*
979-8-88926-175-9   *Hardcover*
979-8-88926-173-5   *Ebook*

To my late aunt, whose unbreakable spirit taught me never to give up.

To my parents, who instilled in me integrity and perseverance.

To my teachers, who kindled my passion for lifelong learning.

You are the reason I am who I am today. This is for you!

Contents

Introduction — 9

I. THE FOUNDATION — 15

1. From Chessboards to Chatbots: A Journey through AI Milestones — 17
2. The AI Revolution: How It's Transforming Business and Society — 29
3. Inside the AI Black Box: A Peek into the Technologies Powering the Magic — 45
4. The Dawn of a New Era: How Generative AI Is Transforming Businesses — 61

II. THE INTEGRATION — 79

5. A Strategic Framework for Enterprise AI Adoption and Integration — 81
6. Fuel for Thought: The Cornerstones and Challenges of Data in AI — 111
7. An Ounce of Prevention: Securing the Competitive Edge of AI — 135
8. The Art of AI Leadership: Building AI Teams — 155
9. AI for Good: Developing AI Responsibly and Ethically — 175

III. TRANSFORMATION: THE ART OF POSSIBILITY — 191

10. Moneyball: How AI Is Revolutionizing Finance — 193

11. The Personal Touch: Crafting AI-Driven Customer Experiences — 213

12. The Art of AI: Redefining Sales And Marketing — 231

13. Operations Overhaul: AI-Driven Transformation — 245

14. AI Transformation in Human Resources: From Recruitment to Retention — 257

15. Building a Better Product: AI Innovation in Engineering — 273

16. AI's Impact on Health-Care Industry — 285

17. AI's Transformative Influence: From Vineyards to Outer Space — 305

IV. THE HORIZON: PREPARING FOR THE FUTURE OF AI — 319

18. The AI Maturity Model: Stages of AI Adoption — 321

19. The Future of Work: AI and Human Collaboration — 339

20. The Road Ahead — 353

Glossary — 369

Acknowledgments — 377

References — 381

Introduction

DROWNING IN PAPERWORK? 996 ANALYSTS? TRY AI!

Nine hundred ninety-six business analysts and weeks of time—that was the issue I faced five years ago. My manager, Scott Herren, tasked me with creating an export compliance function, a monumental challenge that would have required hiring hundreds of business analysts to screen fifty million accounts. To comply with export regulations, we needed to ensure our software was sold to authorized individuals and locations and used within any restrictions imposed by the US and other country laws and regulations, which involved screening potential and existing customers against a complex set of requirements. This daunting task turned into a golden opportunity to leverage AI and robotic process automation (RPA). Unlike other CFOs, Scott was a cross-functional expert, and he fully supported our initiative. Together, we built our own machine learning models using PyTorch. Despite initial skepticism and technical hurdles, my determined team and I developed an ingenious system that automated 99.7 percent of the screening process, saving millions of dollars annually and reducing process time by up to 70 percent.

THE AI REVOLUTION IS HERE!

We stand at the threshold of a transformative epoch poised to redefine business and society. The emergence of artificial intelligence represents a seismic shift. From mastering complex games to unlocking actionable insights from vast datasets, AI has already shown glimpses of its immense potential. AI is rapidly transforming the workforce, with the World Economic Forum estimating that AI could disrupt over 85 million jobs by 2025.[1] We are witnessing AI's rapid evolution, and its implications will be profound, reshaping entire industries and elevating human productivity to unprecedented levels.

THE URGENCY FOR BUSINESS LEADERS

The question facing every business leader is not if AI will impact them but how to harness its power. AI presents an enormous opportunity to reimagine processes, redefine business models, and achieve exponential growth. Mark Cuban emphasized this point during a CNBC interview on March 4, 2024, stating, "If you are a CEO, you can't just say, I am going to get my tech guys to understand it and educate me. You must understand it because it will have a significant impact on every single thing you do."[2] This spirit of continuous reinvention and embracing new possibilities is critical as we navigate the AI-driven future. The AI train is accelerating into a future brimming with possibilities. Climb aboard now or risk being left irrevocably behind. This book provides your ticket to ride.

MY UNIQUE PERSPECTIVE

For over twenty-five years, I have witnessed firsthand AI's rapid evolution from theory to reality. During my graduate studies in computer science in the early 1990s, neural networks first caught my attention. My friend, Ben Carter, and I wrote a game for a psychology lab to evaluate the behavior of smokers using LISP, a programming language for neural networks. At Intel in the late 1990s, I transitioned from Pentium III product manager to senior leadership positions, contributing to the momentum of Moore's law. This era was marked by remarkable strides in computing capabilities, setting the stage for AI's rapid ascent.

By 2017, I was harnessing AI's power in my organization, establishing a data science team that extracted valuable insights from extensive data in background screening and recruitment while driving operational efficiency. Later, AI proved its worth with our export compliance screening. By 2022, the AI hype had already begun. By then, I had deployed multiple AI projects at scale and developed hundreds of use cases that augmented human capabilities to improve productivity and creativity. I firmly believe technology's role is to solve business challenges and augment human capabilities. However, my conversations with CXOs and board members revealed a significant gap between what AI could accomplish and the foundational elements needed for successful AI deployment at scale.

With ChatGPT's release in 2022, AI became a reality for the mainstream public. We have now reached a tipping point where the AI revolution will dwarf even the Industrial Revolution and the internet in its impact.

ADDRESSING PAIN POINTS

I understand the skepticism many business leaders face. Data dilemmas, AI's "black box" lack of transparency, and struggling with where to start—I've been there. That's why *AI Unleashed* provides a practical roadmap to navigate the uncharted waters of AI:

- **Strategic AI Alignment:** Learn how to align AI initiatives with your business objectives, ensuring that every project drives value and supports your strategic goals.
- **Data Governance and Quality:** Discover robust practices for data quality and governance, crucial for reliable AI models and insights.
- **AI Security:** Understand the importance and ways of securing AI assets to protect your competitive edge and ensure business continuity.
- **AI Leadership and Team Building:** Gain insights into building and leading AI teams, focusing on the skills and roles essential for successful AI projects.
- **Ethical AI Development:** Follow ethical guidelines to develop AI responsibly maintaining public trust.

This roadmap is designed to help you overcome initial hurdles and achieve sustainable success in your AI journey, transforming challenges into opportunities for innovation and growth.

ACTIONABLE INSIGHTS

AI Unleashed is tailored for business leaders aiming to explore the potential of AI. You'll gain clarity on how AI could fundamentally transform various functions and your business models, offerings, and value propositions. Use this book as your roadmap to future-proof your organization. While we cover AI concepts and applications, we steer clear of technical intricacies irrelevant from a leadership standpoint. We don't focus on apocalyptic predictions of AI taking over mankind either. Our sights are fixed squarely on one destination—how to understand AI and how it can be a force for good in your business.

A VISION OF THE AI-POWERED FUTURE

AI opens a new world where AI systems swiftly analyze vast datasets, identifying patterns and generating hypotheses at incredible speeds. AI guides researchers to promising avenues, accelerating breakthroughs in medicine, energy, materials science, and more. By augmenting human intellect, AI could enable discoveries in years rather than decades.

Now, picture everyone having a tailored virtual assistant, an intelligent copilot through life. These AI companions provide real-time translation, analyze complex data for decisions, and offer emotional intelligence by detecting moods. Continuously learning and adapting, they become extensions of ourselves, enhancing our capabilities while preserving our individuality.

AI also reshapes urban planning by optimizing transportation networks, energy distribution, and resource management. Intelligent infrastructure adapts dynamically to conditions like traffic and emissions reduction. By identifying revitalization opportunities, AI complements human ingenuity to create ultra-livable smart cities.

However, this AI utopia has a darker side. The efficiency that displaces jobs could cause upheaval if education doesn't keep pace. Deepfakes could blur reality, sowing discord. Malicious actors exploiting AI vulnerabilities could enable nightmarish cyberattacks. We must address these risks to harness AI's immense potential responsibly.

SHAPING THE FUTURE TOGETHER

The challenge is to harness the immense potential of AI while mitigating its risks. We need to ensure responsible development, creating safeguards against bias and misuse. Education must equip us to thrive alongside AI, not be replaced by it. The future with AI is not a preordained script; it's a story we're writing together. Will it be a tale of progress and human flourishing or a cautionary one? The choice is ours, and the decisions we make today will determine the outcomes.

The rise of AI is inevitable. The question is, will you be a leader or a follower? Join me on this journey to unlock the immense potential of AI for your business and humanity. Let's unleash the power of AI together!

SECTION I

THE FOUNDATION

CHAPTER 1

From Chessboards to Chatbots: A Journey through AI Milestones

THE AI ODYSSEY: BEYOND TECHNOLOGY TO PURPOSE AND PEOPLE

When I was in graduate school in the early nineties, Professor Miller, my computer science professor who taught Neural Networks class, told me, "We don't know if we will see the promise of AI materializing anytime soon." We were in the middle of an AI winter then! And we'd had multiple AI winters before.

Our narrative commences in the early twentieth century, marked by the birth of mechanical computers. These initial devices, elementary as they might seem now, were nothing short of groundbreaking. They shattered previous computational limits. The first programmable general-purpose computer, ENIAC, created in the 1940s, stands as a testament to this era—a machine of unparalleled capability for its time, albeit one that occupied an entire room. It cost nearly half a million dollars to build it, equivalent to over six million dollars today.[1]

As the hardware made strides, the software wasn't far behind. The mid-twentieth century heralded the advent of programming languages like FORTRAN and COBOL. The corporate world began to recognize the multifaceted potential of computers, transcending mere arithmetic to encompass data analytics and strategic decision-making.

This served as the backdrop for the emergence of the term "artificial intelligence," which John McCarthy introduced in 1956.[2] McCarthy, who was teaching mathematics at Dartmouth College, organized a two-month workshop to study artificial intelligence in the summer of 1956. This era was filled with hope. The tech community was abuzz with the prospect of machines rivaling human intellect. Endeavors like ELIZA, an early natural language processing computer program simulating a psychotherapist, became the talk of the town by the mid-1960s.

Yet the AI odyssey had its challenges. After initial enthusiasm and investment, it became clear that the technology of the time could not deliver on the promise of "general AI," a form of AI that could perform any intellectual task that a human can do. By the mid-1970s general AI had not materialized in any meaningful way. This led to reduced funding by the US Department of Defense and others and general skepticism about the field's immediate potential. Marvin Minsky, a pioneer in AI and a member of the McCarthy's workshop at Dartmouth, once said, "AI has been brain-dead since the 1970s." This period is called *AI winter*.[3] This sentiment captures the frustration during periods of AI winter, where progress seems to halt. Since then, every decade has seen optimism and crash cycles in AI.

However, the turn of the century saw three developments that catapulted AI to where it is today. First, Moore's law enabled computing power at exponential levels. As a product manager in the late nineties, I was thrilled to bring 1GHz processors built on 250 nanometer technology to the market. For reference, human hair strand is 360x wider than the 250 nm gate length of a transistor. Thanks to Moore's law, today's processors, both CPUs and GPUs are thousands of times more powerful and can handle enormous levels of workloads that were inconceivable twenty years ago.

Second, the internet and the ubiquity of digital devices, ranging from smartphones to Internet of Things (IoT) gadgets spewed enormous amounts of data. The world produces 2.5 quintillion bytes of data every day.[4] That is 2.5 trillion books. If you take those one-inch-thick books and put them around the Earth, they would go around 1,585 times. Lastly, with the advent of machine learning, algorithms have reached an unprecedented level of sophistication.

The Goldilocks moment for AI has arrived!

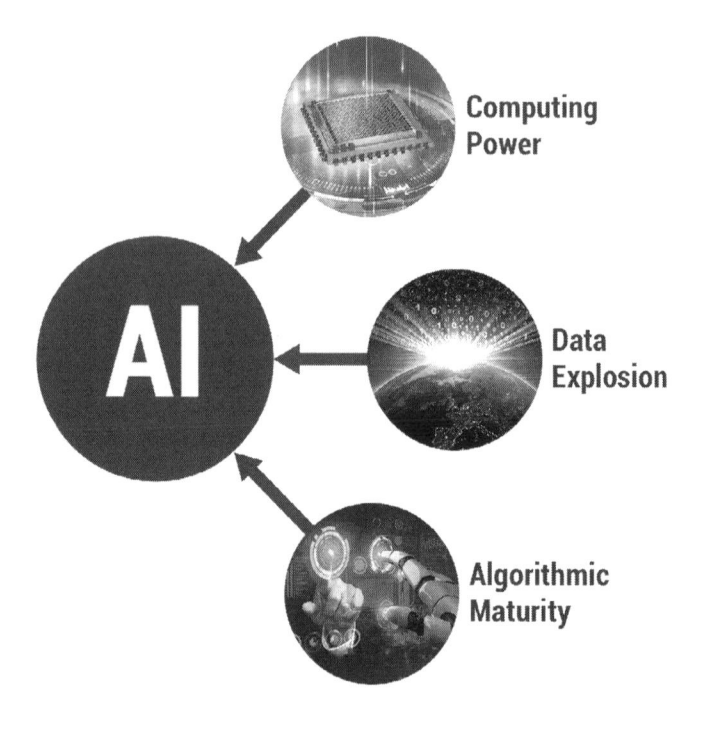

Fig 1. The Goldilocks Moment for AI

DEEP BLUE VERSUS KASPAROV

In the mid-1990s, IBM embarked on a mission to create a machine that could outsmart humanity—at least in chess. The result was Deep Blue, a supercomputer designed specifically to play chess at an elite level. The ultimate test for Deep Blue was to defeat Garry Kasparov, the reigning World Chess Champion and a symbol of human intellect. In 1996, the stage was set for an epic showdown between man and machine. The match took place in Philadelphia, and the world watched with anticipation. Kasparov won, but Deep Blue managed to take a game off him, making it the first computer to ever win a game against a reigning world champion under standard chess tournament conditions.[5]

Not satisfied with the loss, IBM went back to the drawing board to improve Deep Blue. In 1997, a rematch was organized, this time in New York City. The tension was palpable, and the stakes were high, not just for Kasparov or IBM, but for the future of AI. Deep Blue won the six-game match 3.5–2.5, marking the first time a reigning world champion lost a match to a computer under standard chess conditions. The victory was a watershed moment in the field of AI, proving that machines could, in specific domains, outperform human intelligence.

The win had a profound impact on both the world of chess and the field of artificial intelligence. For chess, it marked the beginning of a new era where computers became essential tools for training and analysis. For AI, this significant milestone showcased the potential of machine intelligence to solve complex problems. After the match, Garry Kasparov said, "I lost my fighting spirit."[6] His admission highlighted not just the technical prowess of Deep Blue but also the psychological impact that machines could have on human competitors.

One of the most significant technological advancements in Deep Blue was its parallel processing architecture. The system utilized thirty processors working in tandem, each with its own chess-specific processor. This allowed Deep Blue to evaluate up to two hundred million positions per second, a feat that was unparalleled at the time. Deep Blue used a complex evaluation function to assess the quality of a chess position. This function considered various factors like material balance, piece mobility, and king safety. The function was fine-tuned by a team of chess Grandmasters and IBM engineers, making it incredibly effective at evaluating positions like a human player but at a much faster rate. Deep Blue had an extensive "opening book" containing a variety of opening sequences used by top-level human players. This allowed it to start the game strongly. Additionally, it had a comprehensive endgame database, enabling it to play the final stages of the game with near-perfect accuracy. What set Deep Blue apart was its ability to adapt during the match. Deep Blue couldn't "learn" in the way modern machine learning algorithms do, but the IBM team adjusted its software manually between games to better counter Kasparov's strategies.[7]

ALPHAGO VERSUS SEDOL

In March 2016, an amazing development happened when the world watched as AlphaGo, an AI program developed by Google's DeepMind, faced off against Lee Sedol, a South Korean Go grandmaster. The match consisted of five games and was broadcast globally. The stakes were high, not just for the world of Go but for the field of artificial intelligence as a whole. Go is an ancient board game that is simple to learn but incredibly complex to master. The game has more possible board configurations than there are atoms in the universe.[8] This complexity makes Go a challenging problem for AI, as it's not feasible to compute all possible moves, unlike in simpler games like chess.

AlphaGo used a blend of machine learning techniques to play Go. Monte Carlo tree search (MCTS) algorithm helped AlphaGo explore possible moves and their outcomes, providing a statistical basis for decision-making. AlphaGo used deep neural networks to evaluate board positions and predict the likelihood of winning with certain moves. It had a policy network to sift through possible moves and a value network to estimate the game's outcome. AlphaGo improved its gameplay by competing against different versions of itself, called reinforcement learning, learning from each game to refine its strategy.[9]

AlphaGo won four out of the five games, shocking the Go community and the world at large. The AI's gameplay demonstrated a blend of computational power and, surprisingly, a form of "intuition" in its strategic choices. Many hailed one particular move in the fourth game, Move 37, as exceptionally creative, a move no human would typically consider. "I thought AlphaGo was based on probability calculation, and it was merely a machine. But when I saw this move, I changed my mind. Surely, AlphaGo is creative," said Lee Sedol.[10]

AlphaGo's victory was a milestone for several reasons. The AI demonstrated a form of strategic depth, creativity, and human-like intuition, previously thought to be the exclusive domain of human intelligence. The win validated the effectiveness of machine learning algorithms in solving complex problems, setting the stage for broader applications in various fields. The event sparked global interest in both Go and AI, leading to increased investment and research in machine learning technologies.

AlphaGo's win over Lee Sedol was a landmark event that transcended the domain of board games. It showcased the potential of AI to tackle complex problems and opened the door to a future where AI and humans could collaborate in unprecedented ways. "This has huge potential for using AlphaGo-like technology to find solutions that humans don't necessarily see in other areas," said Demis Hassabis, CEO of DeepMind.[11]

WATERSHED MOMENT: CHATGPT

More details about ChatGPT and Generative AI are covered in chapter 4, but here is a brief background.

Large language models (LLMs) like GPT-3 represent a watershed moment in the development of AI and natural language processing (NLP) capabilities. The launch of ChatGPT in November 2022 was groundbreaking, gaining over one hundred million users within just two months due to its unprecedented conversational fluency and ability to understand context. ChatGPT set a new benchmark by providing more accurate and human-like responses through technological advancements like the transformer architecture, which allowed for improved contextual understanding.

Advancements in NLP paved the way for GPT models. The journey began with narrow rules-based systems to statistical models that calculated word probabilities, and it further expanded to add context and memory with attention mechanisms. However, the transformer architecture in 2017 was the real game-changer, using self-attention to understand relationships between all words simultaneously rather than sequentially. This allowed models to genuinely grasp context and language nuances.

While immensely capable, LLMs like ChatGPT still have limitations in areas like preventing hallucinations of made-up content, efficient retraining from feedback loops, and addressing specialized enterprise needs. The powerful language capabilities of ChatGPT enabled a wide range of potential use cases for businesses—from automating customer service to generating marketing content. As these AI models become more integrated into business operations, implementing human oversight and validation is critical to catch errors and biases while maintaining trust in the outputs. The rise of LLMs sparked both excitement about their potential and acknowledgment that we are still early in harnessing this powerful technology safely and responsibly.

THE OPEN-SOURCE ETHOS IN AI

Open-source tools have been critical in accelerating the growth of AI. The open-source approach promotes the sharing of source code and collaborative development, breaking down barriers to entry and fostering innovation. GitHub and Hugging Face serve as prime examples of this philosophy in action. Both have evolved into global platforms where developers, researchers, and organizations can share code, collaborate on projects, and contribute to the collective knowledge base. GitHub democratizes software development, allowing anyone—from individual developers to large corporations—to contribute to and benefit from the rapidly expanding field of AI.

Two tools that have been instrumental in democratizing AI are TensorFlow and PyTorch.

TensorFlow: Developed by the Google Brain team, TensorFlow is an open-source machine learning framework that allows for the development of neural networks and other machine learning models. It is highly scalable and can be used on a single laptop or scaled across multiple servers. TensorFlow's flexibility and comprehensive ecosystem have made it a go-to tool for both research and production environments.

PyTorch: Originating from Facebook's AI Research lab, PyTorch is another open-source machine learning library that has gained immense popularity. It is particularly known for its dynamic computation graph, which makes it highly flexible and conducive for research and experimentation.

Both TensorFlow and PyTorch are freely available, allowing start-ups and individual developers to access cutting-edge machine learning libraries without the need for significant capital investment. The robust communities around these tools offer extensive documentation, tutorials, and pre-built models, making it easier for newcomers to get started. These frameworks are designed to scale. The open-source nature of these tools encourages a culture of shared innovation. Companies and researchers publish their models and techniques, accelerating the rate of advancement in the field. TensorFlow and PyTorch are highly effective in developing real-world applications—from natural language processing to predictive analytics—thereby enabling even small companies to offer AI-driven solutions. My team used PyTorch to build machine learning models for export compliance screening very effectively as early as 2019.

The transition from rudimentary computing to the intricate world of artificial intelligence stands as a testament to human ambition, unyielding inquisitiveness, and the spirit of innovation. AI's history is a lesson in strategic foresight. While we've made monumental strides, the horizon still beckons.

CHAPTER 2

The AI Revolution: How It's Transforming Business and Society

---

"AI is the new electricity. Just as one hundred years ago electricity transformed industry after industry, AI will now do the same."
— Andrew Ng.

Artificial intelligence (AI) is catalyzing a monumental transformation across industries. The rapid pace of AI innovation and adoption represents a "big bang" moment for the business landscape. According to PwC, AI could contribute up to $15.7 trillion to the global economy by 2030, more than the current output of China and India combined.[1] While these numbers are hard to prove and likely will change in years to come, they represent a magnitude of impact unseen ever before. In recent years, productivity gains have proven to be challenging across companies and economies. That's going to change with AI. This staggering productivity boost underscores AI's potential to reshape entire sectors.

Leading AI capabilities like machine learning, computer vision, and natural language processing are transitioning from research concepts to foundational business pillars. AI is enabling breakthrough innovations across all functions—from supply chain optimization to personalized customer experiences. Legacy firms and disruptive start-ups alike are racing to capitalize on AI's potential. As AI capabilities grow more powerful, responsible oversight is critical. While some tasks can be automated, humans must lead on judgment, creativity, and empathy. Organizations need holistic AI strategies spanning ethics, workforce adaptation, and agile governance.

Companies have started to invest in AI. *AI Magazine* expects the AI market to reach almost $450 billion in 2024 and continue to grow for at least five years.[2] The COVID-19 pandemic further propelled adoption as companies invested in automation to build business resilience. That's just a start. The real opportunity now lies in harnessing its full potential to uplift business and society to unprecedented heights. With strategic vision and human values guiding the way, the future looks bright.

AI is fundamentally reshaping the basis of competition across sectors. Incumbents like Walmart and Toyota are making massive investments in AI to defend their dominance.[3] Meanwhile, digital disruptors like Shopify and Tesla have AI ingrained in their DNA. Start-ups with an AI-first mindset can scale rapidly and contest larger players. Industry boundaries are blurring as AI enables new cross-sector convergence.

According to a McKinsey survey in late 2019, 63 percent of executives believe AI will substantially transform their industry within three years.[4] Companies that fail to adapt risk obsolescence. For every business leader, tech visionary, and forward-thinking executive, understanding this AI paradigm shift isn't just beneficial. It's essential because it's about spotting the next big trend, navigating challenges, and seizing opportunities. Remember, it's not about the technology itself but how AI can change businesses and how we work. And those who harness the power of AI will lead the charge. Mastering AI is becoming essential to stay competitive in the digital age. Viewing it as a peripheral tool rather than a core capability will leave firms behind the curve. Technology analyst Tom Davenport explains, "AI will increasingly be used to provide functionality, differentiate offerings, and remake interfaces. Competing without AI will soon be insurmountable in many businesses."[5]

No matter what your role is, AI will impact it in unprecedented ways in coming years.

RADICAL EFFICIENCY

AI is enabling unprecedented levels of operational optimization and cost efficiency across organizations. AI-driven solutions, ranging from personalized marketing to predictive maintenance, are reshaping cost structures and productivity metrics. Businesses can now forecast demand with uncanny accuracy, streamline inventory, and preempt supply chain disruptions.

In manufacturing, AI techniques like computer vision and machine learning are automating quality assurance and predictive maintenance. Industrial engineering company Siemens leverages AI on factory floors to detect tiny defects in manufactured products. This has reduced their defect rate to nearly zero while increasing throughput.[6]

AI is transforming supply chains as well. UPS uses algorithmic optimization in logistics to save over one hundred million miles annually, thus reducing fuel costs by four hundred million dollars and cutting emissions.[7] Walmart's AI system identifies key sources of waste and offers targeted solutions to its associates.[8]

Customer service is another function experiencing big efficiency gains. Chatbots augmented with natural language processing handle common queries, such as booking confirmations, account balance inquiries, and technical support issues twenty-four-seven without human involvement. Deutsche Telekom automated the resolution of customer issues leveraging their chatbot DT One. AI also drastically reduces customer wait times and service costs by giving the right information from distributed knowledgebase across the company to the customer specialist.[9] According to a study by Capgemini, AI implementations have improved productivity across Research and development (R&D) by 16 percent for major automotive companies.[10] Bear in mind that most companies have not achieved mature AI implementations and at scale. When companies scale AI to enterprise-level, the efficiency gains will likely be significantly higher.

Beyond customer service, AI, armed with predictive analytics and deep learning algorithms, offers a great understanding of consumer preferences. AI can personalize experiences for your customers by knowing their preferences and their history of interactions with your company.

As AI capabilities continue improving, more complex workflows will become automated. Gartner estimates hyperautomation—combining technologies like AI, RPA, and process mining for extreme automation—will enable 40 percent higher operational efficiency.[11] While humans remain integral, AI augments their productivity exponentially. The benefits of AI transcend costs alone. Efficiency unlocks new value streams. As AI handles repetitive tasks, it redirects human effort to innovation and creativity. The synthesis of AI's capacity to handle many of today's human tasks and human ingenuity sets the stage for a new era of productivity and progress.

SPURRING INNOVATION

During my time at Intel, I had the opportunity to interact with the brightest minds in material science and electrical engineering. As the transistors were getting smaller and smaller, we needed to deal with challenges in the design and performance of computer processors. In 2011, our engineers came up with three-dimensional transistors with FinFET technology on a twenty-two-nanometer process node. It took a decade of hard work and ingenuity to reach that milestone and the discovery of high-k dielectric materials and metal gates in transistors enabled the technology.

In late 2023, Google's DeepMind devised an AI program called Graph Networks for Materials Exploration (GNoME) to discover new materials accurately at record speed. Lawrence Berkeley National Laboratory had already synthesized over seven hundred materials from 2.2 million candidates.[12] Contrast this with the ten-year journey to come up with high-k dielectric materials in 2011!

In health care, AI techniques are accelerating drug discovery and improving diagnostic accuracy. Start-up SOPHiA GENETICS uses AI to uncover patterns in genetic data, helping develop targeted therapies ten times faster. AI can also help oncologists in detecting cancer more efficiently from medical scans.[13] AI algorithms were proven to analyze medical images with incredible accuracy in many cases. These algorithms detect anomalies the human eye might miss, aiding in early and accurate diagnoses.

AlphaFold, developed by DeepMind, has made groundbreaking strides in solving the protein-folding problem, a challenge that has stumped scientists for decades. Utilizing advanced machine learning algorithms, AlphaFold predicted the 3D structure of proteins with remarkable accuracy, far surpassing traditional methods. The technology employs deep neural networks to analyze amino acid sequences, using this data to predict how these sequences will fold into intricate 3D shapes.[14] This enormous achievement has far-reaching implications—from drug discovery to understanding diseases—heralding a new era in computational biology and medicine.

Transportation is being reshaped by AI innovations like autonomous vehicles. Google's Waymo has covered over twenty million miles of real-world testing.[15] Its vehicles can now handle complex urban environments without human intervention. I have seen one navigating the streets of San Francisco many times. Tesla is a pioneer in self-driving technology, collecting billions of miles of real-world data to train its AI models.[16] Its vehicles use computer vision, sensor fusion and path planning algorithms to navigate safely on their own.

Think of voice-activated assistants that do more than play your favorite tunes. They manage your smart home or even monitor your health. Media and entertainment companies are leveraging AI for personalized, interactive content. Netflix applies machine learning to tailor recommendations to each subscriber's taste. The company attributes over one billion dollars in savings from churn reduction to its AI recommendation engine.[17] AI also enables immersive media experiences blending the digital and physical worlds.

As AI becomes more advanced, it will catalyze innovations we can scarcely imagine today. Entrepreneurs and researchers will apply AI to unlock breakthroughs across industries and in ways that enhance life. With human creativity augmented by AI, the possibilities are endless.

INSIGHTS TO DRIVE STRATEGIES

AI's prowess in real-time analytics provides invaluable, actionable insights. These revelations empower executives to make strategic pivots, whether in entering new markets or launching innovative products. Consider the example of a retail giant like Walmart. With thousands of stores and an expansive online presence, the company generates a staggering amount of data every day—from customer purchases and inventory levels to market trends and competitor pricing. Traditional data analysis methods would take weeks, if not months, to sift through this data and derive meaningful insights. Enter AI. Walmart employs advanced machine learning algorithms that analyze this data in real time, identifying patterns and trends that human analysts might overlook.[18]

These AI-driven insights empower Walmart's executives to make strategic decisions with unprecedented speed and accuracy. For instance, if the AI system detects a sudden spike in the demand for a particular product in a specific region, the company can quickly adjust its inventory and marketing strategies for that area. This real-time decision-making capability can be the difference between capitalizing on an emerging trend and missing the boat entirely. Moreover, AI's predictive analytics can forecast future market trends, allowing Walmart to make informed decisions about product launches, store openings, or even potential acquisitions.[19] These predictive capabilities enable the company to stay one step ahead of the competition, ensuring long-term success.

AI: THE CREATIVE ASSISTANT

AI is not just a task automator. It provides creative inspiration and assistance. In the visual arts, tools like DALL-E 3 and Midjourney are turning text prompts into novel images artists can build upon. The AI becomes a collaborator in realizing the artist's vision.

In music, AI composition tools help human musicians expand the boundaries of their own creativity. Systems like Amper generate original melodies and harmonies aligned with a musician's desired genre, mood, and style.[20] Rather than replace the artist, the AI acts as a generator of new ideas to inspire the creative process.

For writers, AI narrative tools suggest plot points, analyze tone, and refine story arcs to supplement the human imagination. These tools are not aiming to automate writing entirely; instead, they reduce grind to allow humans to focus on the parts only they can do—establishing the overarching vision and themes.

Of course, we must take care to use AI transparently and ethically. But when implemented thoughtfully, AI becomes a springboard for creativity, not a replacement. It widens human possibilities. As its capabilities grow, we have a remarkable opportunity to transcend previous creative boundaries. The future looks bright for this human-AI creative symbiosis.

Impact Area	Sectors Impacted	Example
Efficiency	Supply Chain, Manufacturing, Customer Service	UPS uses algorithmic optimization to save over 100 million miles annually, reducing fuel costs by $400 million. Siemens uses AI to detect defects in manufacturing, nearly eleminating defect rates.
Innovation	Healthcare, Transportation, Media & Entertainment	Sophia Genetics uses AI for faster drug discovery by analyzing genetic data. Netflix's AI recommendation engine contributes over $1 billion in additional revenue.
Strategic Insights	Retail, Market Analysis	Walmart uses machine learning algorithms to analyze data in real-time, improving inventory management and marketing strategies based on AI-driven insights.
Creative Assistance	Arts, Music, Writing	AI composition tools like Amper help musicians generate original melodies. Narrative tools assist writers in refining story arcs and analyzing tone.

Table 1: AI's Impact from Efficiency to Innovation

While AI is spurring Innovation and creativity and dramatically improving efficiencies, it's also shifting the landscape for organizations. Leaders must actively implement systems to build a dynamic and effective AI-driven company.

For businesses poised to seize the AI opportunity, a well-articulated strategy is a must. This involves identifying AI's unique value propositions, gathering the necessary data, selecting or developing the right tools, and cultivating a culture that embraces AI-driven innovation.

REIMAGINING WORK AND ORGANIZATIONS

The influence of AI extends far beyond technology to radically reimagine human work and organization. By automating routine cognitive and manual tasks, AI is changing the fundamental nature of jobs. A Deloitte study estimates 10 percent of work activities globally could be automated.[21]

This necessitates retraining and upskilling to prepare workers for new roles that focus on socio-emotional, creative, analytical skills rather than rote tasks. Education systems will need to adapt curriculums as well. Some futurists predict the rise of hybrid human-AI teams that combine the complementary strengths of both.

Organization structures and cultures will need to enable human-AI collaboration. Flatter, agile networks may replace rigid hierarchies. Management practices must empower workers and help them adapt to working alongside AI systems. Given AI's fast pace of change, continuous learning and a growth mindset will be critical across teams.

While some roles change, human imagination, empathy, and ethics remain irreplaceable. Prudent leaders recognize AI as an opportunity to remove dull work and uplift their employees. With human values at the helm, AI can play a profoundly positive role in the future of work.

DATA: THE FUEL DRIVING IT ALL

The raw material powering the AI engine is data. Advanced algorithms uncover patterns and insights from massive datasets that would be impossible for humans to analyze. AI's potential hinges on data quality, volume, variety, and veracity.

Leaders must rethink data strategy in the AI era, recognizing it as a core enterprise asset like financial or human capital. This means breaking down data silos, updating infrastructure for analytics at scale, and governing data thoughtfully. For incumbent firms, overcoming legacy data practices is critical.

Equally important is curating high-quality training data to develop accurate AI models. This is challenging when datasets reflect societal biases. Mastering data-driven competitive advantages will separate the AI leaders from laggards. Success lies not in gathering data, but in developing the ability to extract deep, multifaceted insights using AI. Companies that navigate this shift will unleash productivity and innovation at unprecedented levels.

CASE STUDIES: BUSINESSES TRANSFORMED BY AI

Every paradigm shift has its leaders and its pioneers. Brands from diverse industries are not just adapting but leading the charge with AI. Let us dive into their stories.

Personalization: Netflix, once a DVD rental service, has leveraged AI to transform itself into a global streaming giant. Their recommendation engine analyzes viewing habits, ratings, and even the time spent on pausing or rewinding scenes. These insights into user behavior allows Netflix to suggest shows and movies tailored to individual tastes, increasing viewer engagement and subscription retention.[22]

Customer Experience: The global coffeehouse chain, Starbucks, has harnessed AI to enhance the customer experience. Deep Brew, with embedded AI, offers personalized menu recommendations based on factors like weather, time of day, and past orders.[23]

Fraud Detection and Prevention: To safeguard against fraud, American Express, has implemented an AI-driven system that analyzes patterns and flags suspicious activities in real time. This proactive approach has saved millions in potential losses and bolstered customer trust.[24]

AI-enabled Manufacturing: Airbus has integrated AI into its manufacturing processes from predictive maintenance of machinery to AI-driven quality checks. As a result, it has seen a significant reduction in production times and costs. Additionally, AI tools assist Airbus in design optimization, leading to more efficient and sustainable aircraft.[25]

Virtual Try-Ons: Sephora introduced a virtual artist app powered by AI. This app allows customers to try on makeup products virtually before purchasing. Coupled with augmented reality, the AI app gives users a realistic representation, leading to informed purchasing decisions and reduced product returns.[26]

Predictive Maintenance: GE has integrated AI into its operations with their Predix platform, which monitors equipment health in real time. By predicting when machinery is likely to fail, GE can schedule maintenance, reducing downtime and operational costs.[27]

Personalized Playlists: Using AI to analyze listening habits of its users, Spotify is curating personalized playlists with a high degree of accuracy. Whether it's the "Discover Weekly" playlist introducing users to new songs or the "Daily Mix" tailored to individual tastes, AI-driven personalization has been key to Spotify's success.[28]

Demand Forecasting: Zara, the fast-fashion retailer, uses AI to analyze sales data and customer feedback, allowing them to predict fashion trends more accurately. This helps them minimize their unsold inventory and stay ahead of fashion trends.[29]

Creativity Augmented by AI: Known for its suite of creative software, Adobe has integrated AI into tools like Photoshop and Premiere Pro. Their Sensei AI offers features like auto image tagging, smart cropping, and advanced image recognition, thereby reducing manual effort by designers and content creators.[30]

Financial Forecasting: One of the early adopters of AI in the banking sector, JP Morgan, employs AI for tasks ranging from risk assessment to chatbots for customer service.[31]

One notable application is their LOXM trading algorithm, which uses deep learning to execute trades at optimal prices, maximizing profits.[32]

For businesses, AI is a strategic lever. These case studies showcase the transformative power of AI across industries, highlighting the potential for increased efficiency, enhanced customer experiences, and overall business growth. As the digital horizon expands, the businesses that view AI as a strategic partner, rather than just a tool, will be the ones charting a successful course.

THE ROAD AHEAD

The pace of AI innovation is accelerating rapidly. As AI systems become more powerful, their business and societal impact will intensify exponentially.

Mastering AI is becoming essential for competitive survival. Leading companies across sectors are already racing to capitalize on AI's potential. As AI takes on a more central, strategic role, organizations need to build capabilities in responsible AI development, continuous learning, and human-AI collaboration. Managing the ethical risks of bias, transparency and job displacement is crucial. Education and training programs need to prepare workers for adapting to AI.

Regulatory oversight of AI will increase as governments enact policies around safety, privacy, and human impacts. To navigate the uncertainties ahead, leaders need an open mindset focused on maximizing the benefits of AI while mitigating downsides.

The future will involve humans and AI systems working symbiotically to uplift society. It is up to visionary leaders today to chart a course where AI elevates all of humanity. With ethical foundations and human-centered policies, we can build an AI-powered world of new possibilities.

CHAPTER 3

Inside the AI Black Box: A Peek into the Technologies Powering the Magic

"The computing power of GPUs has enabled us to make breakthroughs in everything from medical imaging to conversational AI." —Jensen Huang, CEO of Nvidia

The Goldilocks moment for AI happened with the exponential growth of computing power, availability of vast amounts of data and a radical shift and maturity of the algorithmic techniques. AI's recent leap forward is built on decades of research into key technologies like machine learning, neural networks, natural language processing and more.

IMPACT OF COMPUTING POWER ON AI

In the late 1990s, central processing units (CPUs) were the heart of computing. They were primarily designed for sequential processing tasks. The clock speeds were in the range of hundreds of megahertz, and they had only a single core. Graphics processing units (GPUs) existed but were focused on rendering graphics for video games. As we entered the new millennium, CPUs started to evolve.

When I was a product manager of Pentium *III*, my team and I at Intel were thrilled to ship the first 1GHz CPU built on 250 nm process technology in March of 2000. A few years later, Intel and AMD introduced multi-core and multi-threaded processors. Around the same time, GPUs began to evolve. Nvidia introduced Compute Unified Device Architecture (CUDA) in 2006, allowing developers to use GPUs for general-purpose computing.[1] The 2010s saw Nvidia and AMD started producing GPUs optimized for AI and machine learning tasks. These GPUs had thousands of smaller cores designed for parallel processing, making them ideal for tasks like image recognition and natural language processing.

For comparison purposes, a 5 GHz sixteen-core CPU today is about eighty times faster than a 1 GHz single-core CPU two decades ago. A modern GPU with four thousand cores running at an effective speed of 1.5 GHz is six thousand times faster than a 1 GHz single-core CPU for tasks that can be parallelized. This incredible speed-up enables today's advancements in AI, data analytics, and scientific research. The difference in computational speed and efficiency is akin to going from a bicycle to a hyperloop.

Today, we have CPUs and GPUs specifically designed for AI tasks. Companies like Google have even developed custom chips like tensor processing units (TPUs) optimized for machine learning. These chips are designed to accelerate tensor computations, the core operations in deep learning algorithms. The advancements in CPUs and GPUs have had a profound impact on enabling machine learning and AI.

MACHINE LEARNING

At its core, AI relies on algorithms—sets of instructions for solving problems in an efficient, methodical manner. Based on probability theory AI scientists developed machine learning (ML). For two decades, machine learning suffered from lack of good amounts of data and computing power to run the algorithms fast. But by the late nineties, that started to change thanks to the explosion of data from the internet and devices along Moore's law powering CPUs and GPUs.

Machine learning takes a data-driven approach. Instead of hard-coded rules, algorithms are trained using statistical models built from training data. Think of machine learning as teaching computers to learn from experience, much like humans do. ML algorithms enable computers to improve at tasks with experience and data. A machine learning model examines inputs and outputs to find patterns, progressively enhancing its performance on a task like image recognition or speech transcription. Picture it as nurturing a young mind. You celebrate the right choices and guide it through the missteps. Based on statistics and probability theory, these models identify subtle correlations within massive datasets that would be impossible for humans to discern.

Prior to machine learning, AI systems relied on hand-coded rule-based programming. Experts manually defined exhaustive sets of rules and logic to enable tasks like playing chess or solving mathematical problems. However, this approach had key limitations. They were brittle—rule-based systems that broke easily outside narrowly predefined conditions. They lacked flexibility to handle real-world complexity and nuance. They were also labor intensive as encoding rules and logic manually requires vast expert effort and is prone to human error. Updating systems with new rules was cumbersome, and rule-based systems focused on narrow domains and struggled to extend to broader contexts.

Machine learning offers three key advantages—adaptability, scalability, and broad applicability. Systems continuously improve by learning patterns from new data. This enables flexibility and resilience. Once we train algorithms, they can be applied widely without rewritten rules. Systems can be updated by retraining models. Machine learning techniques apply across diverse domains from computer vision to NLP.

Let's dive a bit deeper into the training of machine learning systems.

SUPERVISED LEARNING

The supervised learning concept is an important part of AI. Here, the system learns from a map, where every input has a known output. Supervised learning needs to have two sets of data—positive and negative examples. Humans label the data into those categories. These two sets, called the training set, are used to train the AI. Once we train the system, we use test data to check the performance of the AI system.

You are a business executive who wants to improve customer satisfaction (CSAT) scores. You decide to use AI to predict which customer interactions are likely to result in low CSAT scores so you can proactively address issues. In this scenario, the "map" would be a historical dataset of customer interactions. Each input could be a set of variables like the duration of customer service calls, the number of issues resolved, and the type of service provided. The output is the CSAT score—either "High" or "Low." You gather thousands of past customer interactions, already labeled with "High" or "Low" CSAT scores by your customer service team. This is your training set. The AI system learns from this training set. It identifies patterns—like shorter call durations and more issues resolved often lead to "High" CSAT scores.

After we train the AI, we use a separate set of customer interactions (the test data) to see how well the system can predict CSAT scores. By doing this, you've just employed supervised learning to potentially improve your customer satisfaction rates. The AI system has been "supervised" because it learned from a dataset where the outcome (CSAT score) was already known.

While supervised learning is powerful, it often requires extensive labor and resources. Image classification requires millions of labeled images while NLP datasets can range from thousands of sentences to entire corpuses, and customer behavior prediction might need millions of customer records. Human experts must manually label thousands or even millions of data points to train supervised models. For example, autonomous vehicles are trained on countless labeled images of roads, signs, and objects.

Producing this labeled data is expensive and time-consuming. The process also requires meticulous quality control by human labelers with domain expertise, further adding time and cost.[2] In certain specialized fields like medicine, sufficiently large, labeled datasets may not even exist yet. The dependence on scarce labeled data creates a bottleneck, as the available labeled data may not adequately represent all possible real-world scenarios. Even large organizations struggle with the complexity of managing large-scale labeling workflows across huge, ever-growing datasets required for cutting-edge AI. For many, high-quality labeled data remains the key obstacle in effectively leveraging supervised learning.

UNSUPERVISED LEARNING

Unsupervised learning is another crucial aspect of AI. Unlike supervised learning, where the system learns from a labeled "map," unsupervised learning operates without a guide. In this approach, we give the AI system a dataset without explicit instructions on what to do with it. The system tries to learn the patterns and the structure from the data without any supervision. The primary goal is often to identify clusters or groups within the data. Once the system identifies these clusters, they can be used for various applications, including customer segmentation, anomaly detection, and recommendation systems.

Let us say you're looking to expand your product line. You have a wealth of customer data but are not sure what products might resonate with your audience. You decide to use AI to help you understand your customer base better. You feed the system a dataset that includes various customer attributes like age, purchase history, location, and browsing behavior. The data has no labels; the AI system has no idea what it's supposed to find. The AI goes to work, analyzing the data to identify clusters of similar customers. After the analysis, you find distinct groups within your customer base have similar behaviors and preferences. One group might be young professionals interested in tech gadgets while another could be parents who frequently purchase children's clothing. You didn't provide the AI with any specific outcome to look for; it found these clusters on its own. Now, you can tailor your new product line to meet the unique needs of these different customer segments.

Unsupervised learning empowers businesses to uncover hidden customer segments from data autonomously. By identifying distinct customer profiles and preferences, companies can tailor products to meet specific needs, driving innovation and growth in an ever-evolving market without relying on predefined labels or instructions.

REINFORCEMENT LEARNING

Reinforcement learning is another cornerstone in the AI landscape. Unlike supervised learning, which relies on labeled data, reinforcement learning is more like learning by trial and error. The AI system, often referred to as an "agent," interacts with an environment to achieve a specific goal or maximize some notion of cumulative reward.

In reinforcement learning, a robot dog learns not from a pre-labeled dataset but through real-world random interactions. When given a command like "fetch," the robot dog embarks on a journey of trial and error, exploring various actions—such as move forward or backward, grab the ball, or bring it back—to see which one yields a reward. If fetching the ball garners a positive response, the dog's internal policy adjusts to favor that action in the future.

For instance, if the dog initially fails to move upon hearing "fetch," it receives no reward and learns nothing. However, when it eventually decides to chase the ball and succeeds, it earns a reward, reinforcing the idea that fetching is the optimal action. Over time, this process of exploration and policy adjustment helps the robot dog become proficient at fetching, as it learns to maximize rewards through its actions.

The AI agent starts with little to no information about the environment. It takes actions and observes the outcomes, receiving either rewards or penalties based on how well those actions align with the goal. Over time, the agent learns to make better decisions that maximize the cumulative reward.

Here is how it works in a business case. You are focused on optimizing your supply chain and minimizing costs while ensuring timely deliveries. In this context, the "environment" is your supply chain, and the "agent" is the AI system you deploy to manage it. The "actions" could be various logistics decisions like which supplier to use, which route to take, or how much inventory to stock. The "reward" is a measure of how well the supply chain is optimized, perhaps based on factors like cost savings, delivery times, and customer satisfaction.

The system continually refines its strategy, adapting to new information and challenges to optimize the supply chain operations. By employing reinforcement learning, you're not just making decisions based on historical data; you're enabling your AI system to adapt and improve in real time. This approach can be incredibly powerful for complex, dynamic systems like supply chains, where conditions can change rapidly, and the optimal decision is not always obvious.

```
                        Machine
                        Learning
            ┌──────────────┼──────────────┐
      Supervised      Unsupervised    Reinforcement
       Learning         Learning         Learning
           │                │               │
        Labeled          Unlabeled         Agent-
         Data              Data         Environment
                                         Interaction
        ┌──┴──┐          ┌──┴──┐         ┌───┴───┐
  Classification Regression Clustering Association Policy Reward
```

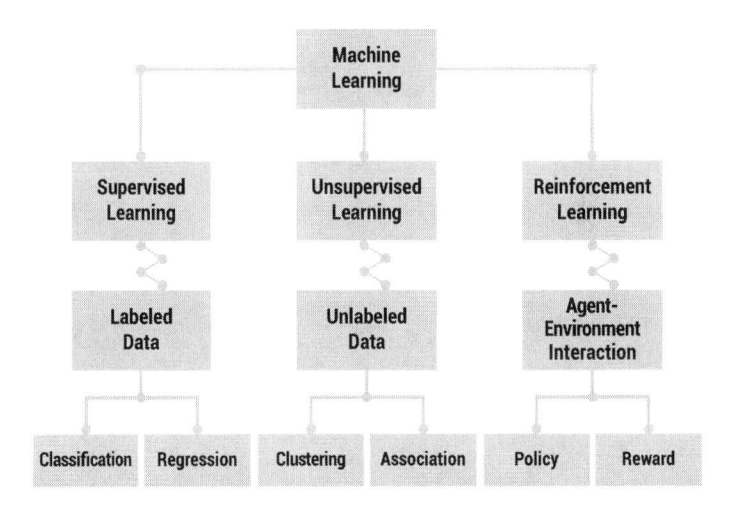

Fig 2: Types of Training Machine Learning Systems

NEURAL NETWORKS

Neural networks are a specific type of machine learning model. Sometimes called artificial neural networks (ANNs), they mimic how neurons in the brain signal one another. They are the architects of pattern recognition. Elements in neural networks are simulated neurons with inputs and outputs along with one or more hidden layers. Using multiple layers of interconnected nodes, they can model extremely complex relationships between inputs and outputs. The nodes operate in parallel, allowing very rapid information processing.

Neural networks underpin recent AI breakthroughs in computer vision, natural language processing, and more. They translate the world's data, much like our senses do for us.

To illustrate neural networks by the concept of identifying animals, let's say we want to build an AI system that can differentiate between various animals like cats, dogs, and sheep. The input layer of the neural network would receive the data about the animal, like an image or audio clip. The nodes in the input layer would detect basic features like edges or tones. The hidden layers would spot more complex features, like facial shapes, textures, four legs versus two legs, etc. Each node combines inputs from the previous layer to extract higher-level information. The output layer classifies the animal based on the derived features. If the hidden layers detect whiskers, pointy ears and a soft purr, the output layer recognizes: "This is a cat!" In other words, the input layer receives the raw data, hidden layers extract relevant features and patterns, and the output layer uses those learned features to make a final classification or prediction.

By having multiple hidden layers in between, neural networks can learn very complex concepts like identifying diverse animals in images or understanding speech. The network builds an intuitive understanding through exposing it to more training examples over time.

DEEP LEARNING

Deep learning, a child of machine learning, is a multilayered neural network. Multiple hidden layers exist between the input and output layers. This setup mirrors the human brain's capacity for thought, reflection, and dreaming, adept at navigating through extensive data realms. A single-layered network might give you a sketch, but the multilayered ones paint the full picture.

Backpropagation stands as a cornerstone in the training of artificial neural networks (ANNs), a concept that has been pivotal in the field of deep learning. Conceptualized by Geoffrey Hinton and others, backpropagation is a supervised learning algorithm, relying on the availability of labeled data to guide its training process.[3] The essence of backpropagation lies in its methodical approach to error correction. It begins at the output layer of the ANN and meticulously works its way backward. Throughout this journey, backpropagation adjusts the weights of the connections between neurons, effectively tuning the network. This adjustment is based on the magnitude of error in the predictions the network makes.

By iteratively minimizing these errors, the ANN gradually improves its ability to make accurate predictions on future data, embodying a learning process that mirrors the incremental nature of human learning. This technique not only embodies the rigor of Hinton's pioneering work but also encapsulates the intricate interplay between data, error, and learning that is fundamental to the evolution of artificial intelligence.

Feature	Reinforcement Learning	Backpropagation
Type of learning	Unsupervised	Supervised
Data requirements	No labeled data	Labeled data
Goal	Learn a policy that maximizes expected reward	Learn to make accurate predictions
Training method	Trial and error	Gradient descent
Applications	Robotics, game playing, control systems	Image recognition, natural language processing, speech recognition

Table 2: Reinforcement Learning vs Backpropagation

In 2012, a team at Google AI developed a sophisticated multilayer neural network endowed with over a billion weights and exposed it to millions of random YouTube videos for processing.[4] The neural network adjusted the weights as it watched the videos. At the end of the training, it identified cat images successfully. This is an example of multilayered neural network or deep learning. Consider the marvels around us: Siri's voice, Spotify's song suggestions, or even the uncanny accuracy of your phone camera. Deep learning is behind all of them.

NATURAL LANGUAGE PROCESSING

Natural language processing (NLP) is a type of machine learning that enables computers to process human language in the form of text or voice data and to "understand" its full meaning, complete with the speaker or writer's intent and sentiment. NLP is a complex and challenging field that draws from many disciplines, such as linguistics, computer science, psychology, and neuroscience.

NLP transcends mere word decoding, delving into the emotions, context, and the very essence of human communication. Some of the common tasks NLP performs include converting spoken words into text data, extracting meaning and intent from natural language data, producing natural language data from structured or unstructured information, translating natural language data from one language to another, determining the attitude or emotion of a speaker or writer, creating a concise summary of a large text document, and finding relevant answers to natural language queries.

Given these tasks, NLP has applications in many domains from education, health care, and entertainment to social media. Some examples of NLP-powered systems are:

- **Digital assistants** are software agents that can perform tasks or services for users based on voice or text commands, such as Siri, Alexa, or Cortana.
- **Chatbots** are conversational agents that can interact with users via natural language, such as customer service bots or virtual assistants.

- **Search engines** can retrieve relevant information from large collections of data based on natural language queries. Examples are Google, Bing, or Perplexity.
- **Text analytics** systems can extract insights and patterns from large volumes of text data, such as sentiment analysis, topic modeling, or keyword extraction.
- **Speech synthesizers** can generate human-like speech from text data, such as text-to-speech or voice cloning.

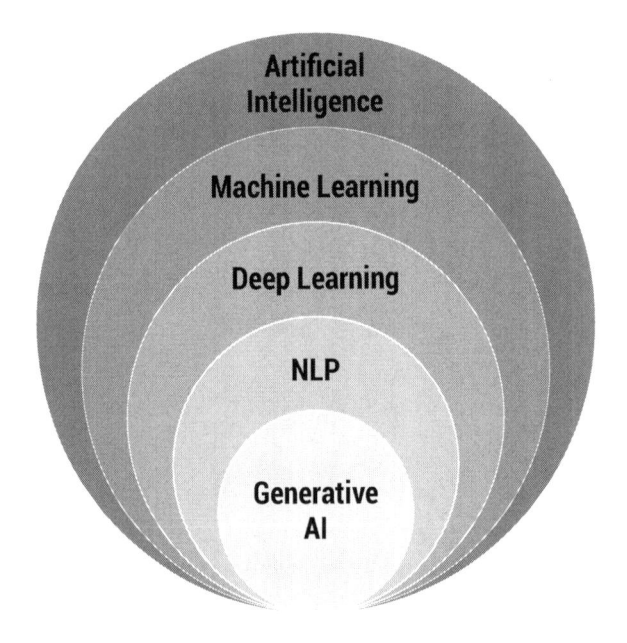

Fig 3: Types of Artificial Intelligence

NLP is a rapidly evolving field that is constantly developing new techniques and models to improve the performance and accuracy of natural language systems. Some of the recent advances in NLP are transformer models, a type of neural network architecture that uses attention mechanisms to capture long-range dependencies and context in natural language data, such as BERT, GPT-3, or T5. Generative pretrained transformers (GPT) are a family of neural network models, and we will cover them in detail in the next chapter.

For all its sophistication, AI remains a human endeavor. It springs from our insight, imagination, and aspiration to build tools that augment our capabilities. As computers grow increasingly capable of flexibly applying knowledge, the torch is passed to human ingenuity to keep exploring new frontiers. The symbiosis between computational prowess and human creativity sets the stage for AI systems that can work collaboratively with people across every field and industry. With open and curious minds, inspired hearts and a steadfast moral compass, our AI journey has only just begun.

CHAPTER 4

The Dawn of a New Era: How Generative AI Is Transforming Businesses

---

In November 2022, the AI world witnessed a groundbreaking moment with the introduction of ChatGPT. It gained one hundred million users within two months of its launch, making it the fastest adopted consumer app.[1] ChatGPT brought an unprecedented level of conversational fluency and contextual understanding, setting a new benchmark for NLP models.

Unlike its predecessors, ChatGPT understands context better, provides more accurate responses, and engages in a more human-like dialogue. A series of technological advancements and fine-tuning made ChatGPT into a sophisticated conversational agent. The launch set the media world ablaze while the AI community, brimming with anticipation, meticulously unraveled its prowess, subjecting it to a diverse array of trials to gauge its practical potential.

The introduction of ChatGPT marked a significant milestone in the journey toward achieving truly conversational AI. For the first time, we had an AI model that could not only understand and generate text but could also engage in meaningful dialogue over extended interactions. This opened a plethora of opportunities for businesses in customer service, content generation, and even decision-making support systems. ChatGPT became the new gold standard, challenging other players in the AI field to catch up or become obsolete.

The market's response to ChatGPT was overwhelmingly positive. Businesses were quick to recognize its potential for enhancing customer experience and operational efficiency. Early adopters ranged from tech start-ups to Fortune 500 companies, all eager to integrate ChatGPT.

A report by McKinsey estimates Generative AI could deliver $2.6–4.4 trillion per year in value across sixty-three use cases analyzed, adding 15–40 percent to the impact of overall AI. While actual numbers may be very different from these, it's clear generative AI can drive major productivity gains through automation, content creation, data analysis, and personalization.[2]

WHAT IS GPT?

GPT stands for generative pre-trained transformer. It's a type of artificial intelligence model designed for natural language processing tasks. The name itself provides a hint about its structure and functionality:

- **Generative:** The model can generate coherent, diverse, and contextually relevant text over extended passages.
- **Pre-trained:** Before being fine-tuned for specific tasks, the model is trained on vast amounts of text, audio, and video data to understand language structure, semantics, and context.
- **Transformer:** This refers to the underlying architecture used by the model, which allows it to handle sequential data, making it particularly effective for language-based tasks.

A series of innovations in natural language processing paved the way for GPT. In early language models, computers could only understand words in isolation and perform narrow predefined tasks. They lacked true language comprehension. Then came statistical models that calculated the probability of certain words following others. This was an improvement but still limited. The next wave brought context and memory. Models could recall previous words to better predict the next word in a sentence. Attention mechanisms helped them focus on the most relevant words.

But the real breakthrough was the transformer architecture in 2017. Instead of processing words sequentially, it uses self-attention to understand relationships between all words simultaneously.[3] This self-attention mechanism was a momentous change. It could process words in a sentence all at once, rather than one at a time, making it incredibly fast and efficient. Think of it as having a team of experts who can all work on various parts of a problem at the same time and then instantly combine their insights to give you the best answer. This architecture gave models a much deeper capacity to grasp context and nuanced language.

The transformer became the foundation for GPT models. By analyzing vast text corpora, GPT achieved unprecedented fluency in generating human-like text. In essence, self-attention allowed models to transcend narrow symbol-processing and enabled genuine language understanding. This shift from rigid rules to dynamic understanding unlocked AI's potential for natural language tasks.

Here's a simplified explanation of how GPT works:

- **Tokenization:** The AI breaks down a block of text into tokens, which are essentially pieces of words the model can understand and process. Each language model uses slightly different tokenization, but broadly speaking we can assume seventy-five words are equal to one hundred tokens. See glossary for additional details.
- **Word Embeddings:** Each token is then converted into a vector, known as a word embedding, which quantifies the linguistic features of a word. These embeddings help the model understand the meaning of words in context.
- **Self-Attention Mechanism:** The transformer employs a self-attention mechanism that allows the model to evaluate the importance of other tokens in understanding the meaning of a word within a sentence.
- **Contextual Understanding:** By processing all the words in a sentence or text block at once, the model can generate or translate text with a high degree of accuracy, recognizing the nuances of language use.

EVOLUTION OF GPT MODELS

Generative AI aims to mimic human creativity and imagination, allowing for the generation of novel outputs including text, code, images, music, and other formats. OpenAI's ChatGPT has emerged as a groundbreaking chatbot rooted in large language model (LLM) technology. LLMs are a specific type of generative AI that focuses on processing and generating text.

ChatGPT, Claude, and Gemini have been fine-tuned to excel in open-ended dialogues. The techniques behind their conversational abilities include reinforcement learning from human feedback (RLHF) and constitutional AI training. RLHF involves humans providing feedback to refine the model's outputs while constitutional AI aims to instill the model with key capabilities like truthfulness and avoiding harmful outputs.

Generative AI is in constant flux, with industry giants like OpenAI, Anthropic, and Google at the helm of pioneering highly sophisticated language models. These entities are setting new benchmarks in AI's capabilities, efficiency, and commitment to generating safer outcomes. A look at some of the forefront models reveals:

- **GPT:** GPT-4o is the latest model from OpenAI, the company behind ChatGPT. Some estimated the number of parameters in GPT-4 model were over a trillion compared to the original GPT with 110+ million.[4] OpenAI hasn't revealed the actual number. The volume of parameters determines the capacity of a neural network to absorb information.

- **Claude:** Claude-3 boasts several enhancements, such as the ability to process an astounding 100k+ tokens per prompt and generate outputs that span thousands of words.[5] The model has been refined for accurate and sensitive responses while minimizing the risk of harmful or offensive outputs.
- **Gemini:** Gemini 1.5, developed by Google DeepMind, represents a significant leap forward in generative AI. This model integrates advanced reasoning capabilities with enhanced natural language understanding, making it adept at more complex and nuanced interactions. Gemini 1.5 boasts a dramatically extended context window, capable of processing up to one million tokens, enabling it to handle vast amounts of data efficiently.[6]
- **Open Source:** For companies looking for Open Source LLMs, projects like Hugging Face's Transformers and EleutherAI's GPT-NeoX are democratizing access to state-of-the-art language models. The latest versions of Llama and Mixtral are excellent open-source alternatives.

Generative pre-trained transformer models have emerged as a breakthrough, allowing large language models to generate human-like text with remarkable fluency. As generative AI rapidly advances, these language models are expanding the possibilities of human-machine interaction and AI.

HOW GPT CAN HELP BUSINESSES

My good friend, Dave Cruse, took a sip of his latte as he recounted his recent experience to me over coffee in Portland, Oregon. As a solopreneur managing his own rental business, he had been leveraging ChatGPT to streamline various tasks—from automating emails to tenants and setting up alerts for maintenance requests. He faced a particularly challenging situation when one of his tenants unexpectedly vacated the rental unit in the middle of their lease term. Uncertain of the best course of action, Dave turned to ChatGPT, which provided him with a comprehensive overview of his options and helped him draft a well-crafted letter to the errant tenant. While the AI assistant didn't replace the need for legal counsel, it equipped Dave with the necessary information to take the next step in resolving the issue, allowing him to navigate the complex situation with greater confidence and preparedness.

While the use cases continue to grow for generative AI, here are some ways it can help businesses.

PROCESS AUTOMATION

Generative AI tools like Claude or ChatGPT can automate a variety of tasks—from replying to customer emails to summarizing lengthy documents. This not only boosts efficiency but also frees up human capital for more strategic activities. Let's say your accounts payable department is swamped with invoices. GPT can be programmed to read, categorize, and even approve or flag these invoices based on predefined criteria. It's like having an extra set of hands work tirelessly, ensuring payments are made on time and anomalies are caught early.

AUTOMATING CUSTOMER SERVICE

Your customer service team is a busy airport control tower, juggling multiple flights (customer queries) at the same time. GPT acts like an advanced autopilot system, handling routine flights so your human controllers can focus on more complex tasks. It can answer frequently asked questions, guide users through processes, and even escalate issues to human agents when needed. The result? Faster response times, happier customers, and a more efficient customer service operation.

CONTENT GENERATION AND CURATION

Generative AI can create natural language text, code, images, videos, music and more tailored to specified topics or styles. This has applications in marketing, media, education and entertainment. GPT is your in-house content creator that never sleeps. Whether you need blog posts, social media updates, or product descriptions, GPT can generate them quickly. It's like having a twenty-four-seven newsroom that can churn out quality content tailored to your audience. Plus, it can curate content by summarizing long articles, recommending relevant pieces, and even generating meta-descriptions for SEO.

DATA ANALYSIS AND INSIGHTS

Drilling through data to find insights can be a complex task. GPT can be your mini data scientist, sifting through large sets of data to identify trends, anomalies, and opportunities. It functions as a super-smart detective who can quickly connect the dots, providing you with actionable insights that can drive business decisions. Claude, ChatGPT, and Gemini can quickly analyze large datasets, revealing valuable insights useful in various industries.

PERSONALIZATION

Recommendation engines, targeted ads, and custom experiences can be powered by generative AI to tailor content and services to each user. An ed-tech platform can use GPT to analyze a student's past performance and learning style, subsequently generating a personalized learning path. An online retail platform can utilize GPT to analyze user behavior and preferences, offering personalized product recommendations and targeted promotions.

ChatGPT can function as a personal tutor, providing students with instant clarifications, explanations, and additional resources tailored to their specific queries and learning pace. A friend's son who is studying pharmacy recently told me that he uploads his class notes into ChatGPT and it creates quizzes to test his knowledge and has been very helpful.

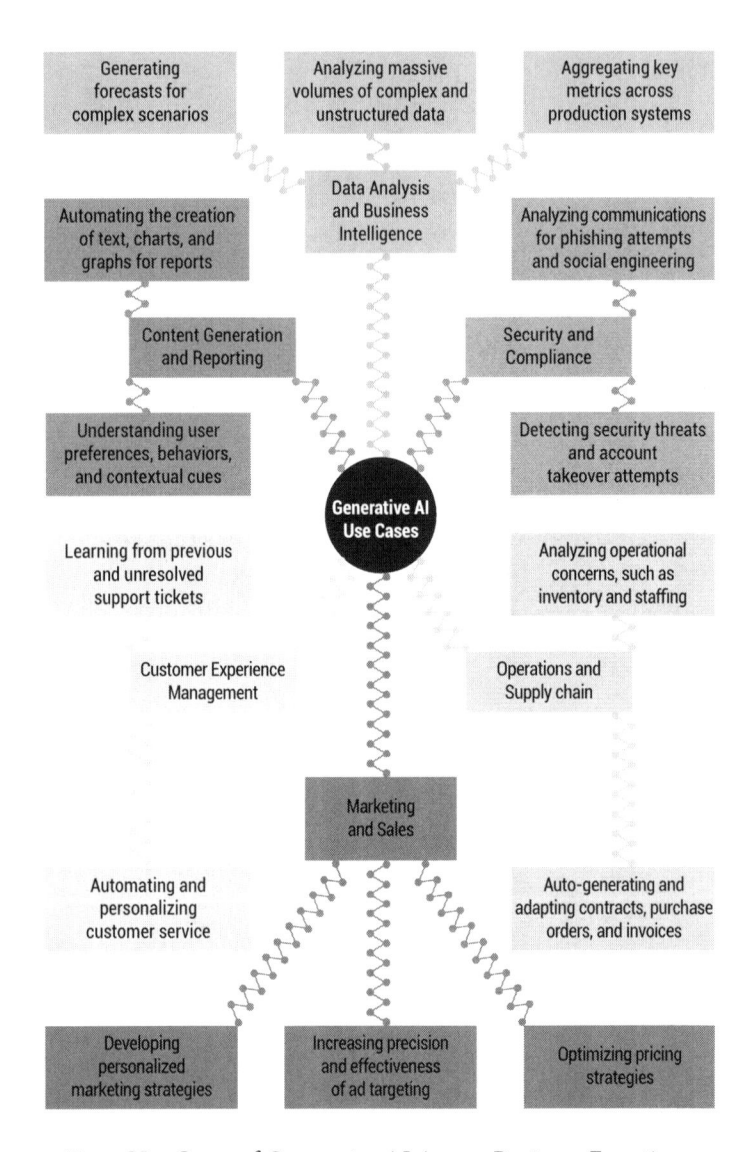

Fig 4: Use Cases of Generative AI Across Business Functions

MAKING APPS INTELLIGENT WITH GENERATIVE AI

Integrating generative AI into applications ushers in a new era of intelligence, transforming them from static tools into dynamic assistants. By infusing AI capabilities, existing applications evolve, gaining the ability to understand and respond to natural language queries, offering a more intuitive user experience. Developers can craft bespoke copilots tailored to specific professional needs, enhancing productivity and decision-making. This AI infusion also paves the way for the creation of novel applications such as AI-powered code editor or personalized learning platforms designed from the ground up with AI at their core, capable of delivering personalized experiences based on data insights and learning from interactions to refine their functionality over time.

Beyond enhancing user engagement, this AI-driven approach significantly accelerates the development cycle, enabling rapid deployment of new features and ensuring a swifter journey from concept to market. In essence, generative AI serves as the catalyst that not only smartens applications but also democratizes innovation, allowing businesses to stay agile and competitive in the digital age.

While generative AI, such as GPT, offers immense potential for business innovation through process automation and intelligent content generation, it's important to pivot and consider the challenges. As we transition to the integration of large language models (LLMs) in enterprise settings, we must be mindful of their limitations.

LIMITATIONS OF LLMS IN ENTERPRISE APPLICATIONS

Large language models (LLMs) have shown remarkable capabilities in various applications, but their deployment in enterprise settings reveals several limitations that organizations must navigate. Understanding these challenges is crucial for businesses considering integrating LLMs into their operations.

HALLUCINATIONS

A big challenge with LLMs today is hallucinations, manifesting as the generation of plausible yet unfounded content. This phenomenon is particularly problematic in enterprise applications where precision is nonnegotiable. Suppose a financial forecast is sprinkled with convincing but imaginary numbers, or a legal document laced with concocted precedents. The root of this issue lies in the LLMs' training on diverse datasets that do not equip the models with the discernment to distinguish fact from fiction. Mitigating these creative liberties requires a blend of enhanced training protocols aimed at sharpening the models' grasp on reliability and the incorporation of validation layers to sieve out the chaff of hallucinations. For enterprises, this means not just harnessing the generative power of LLMs but also curating a symbiosis between AI's creative potential and human expertise to ensure the integrity of the output.

DEPLOYMENT MATURITY

The deployment of LLMs in real-world enterprise applications is still in its nascent stage. Despite the high volume of experimentation with LLMs, based on my discussions with many board members and C-suite executives, very few well-documented cases exist concerning successful integration into business applications. This lack of proven deployment examples can make enterprises hesitant to commit resources to integrate these models into their workflows.

FEEDBACK LOOP EFFICIENCY

For LLMs to evolve and improve, a robust feedback loop is essential. Feedback loop refers to the system's ability to learn from interactions, adapt to new information, and improve over time based on user feedback. An efficient loop allows LLMs to continuously refine their outputs, making them more accurate, relevant, and tailored to specific user needs or business requirements.

Consider an LLM deployed in a customer service application to handle user queries. Initially, the model might provide generic responses that don't fully address specific user concerns. Through a feedback loop, users or supervising employees can flag inadequate responses, providing correct answers or guiding the model toward better responses. The LLM then incorporates this feedback, learning from the corrections and improving its future responses. However, if this process of collecting, processing, and integrating feedback into the model is cumbersome or slow, it hampers the LLM's ability to adapt and evolve.

VALUE PROPOSITION FOR ENTERPRISES

Even though LLMs have a lot of advanced features, making them work for a specific business goal often needs a lot of extra work and changes. For example, a big, general-purpose LLM might not be ready to handle a company's unique customer service questions right away. It might take a lot of time and effort to adjust the model to understand and respond to those specific questions correctly, making businesses wonder if it's worth the effort in the first place.

DATA PRIVACY AND CONTROL

Keeping information safe and under control is very important for businesses, especially when it involves private or special company data. Open-source LLMs let companies have more control over their data, but making sure this data stays private and follows the rules needs careful handling. For example, a business might be able to change an open-source LLM to better fit its needs, but it also has to work hard to make sure this doesn't lead to private information getting out. So, businesses need to think carefully about whether the advantages of customizing these models are worth the extra effort needed to keep data safe.

COMPLEXITY AND GOVERNANCE

The deployment of open-source LLMs involves navigating complex governance, licensing, and support challenges. These factors can introduce additional layers of complexity to the deployment process, making it more daunting for enterprises to adopt these models compared to using more straightforward, albeit potentially less flexible, API services from providers like OpenAI.

LANGUAGE AND CONTEXTUAL LIMITATIONS

Language and contextual limitations present a significant hurdle in the global application of LLMs. In my testing with models like ChatGPT, Claude, and Gemini for translating into languages such as French, Italian, Mandarin, Telugu, and Hindi, I observed while translations into French, Italian, and Mandarin were impressively accurate, the models struggled with capturing the contextual and cultural nuances in other languages. This inconsistency can pose challenges for international businesses, necessitating further refinement and customization of these models to ensure they can accurately interpret and respond within varied cultural and linguistic frameworks.

HYBRID DEPLOYMENT CHALLENGES

Using a mix of open-source and closed LLMs can make things tricky for businesses. This approach lets them use the best parts of both types of LLMs, but it also means dealing with the challenge of making different systems work together smoothly. For instance, a company might use an open-source LLM for its flexibility in one project and a closed LLM for its robust features in another. However, making sure these two systems can talk to each other and work well together can make their AI strategy more complex.

COST AND RESOURCE IMPLICATIONS

Although open-source LLMs might offer cost advantages in the long run, especially for companies with existing infrastructure, the initial investment in customization, deployment, and ongoing management can be significant. Enterprises must carefully consider these costs against the expected benefits and value addition from integrating LLMs into their operations.

While LLMs hold great promise for transforming enterprise applications, organizations must carefully consider these limitations and challenges. A thoughtful approach to integration, emphasizing customization, data privacy, and strategic deployment, can help enterprises navigate these challenges and harness the full potential of LLMs.

THE DOUBLE-EDGED SWORD OF AI: TRUST BUT VERIFY

Companies are increasingly considering using AI to automate tasks—from code generation to summarizing documents. However, the technology is far from infallible. It's prone to errors, biases, and even hallucinations, where the AI produces inaccurate or nonsensical outputs.

The importance of keeping a "human in the loop" to validate AI-generated content cannot be overemphasized. Yet we see a gap in execution. According to a survey by the staffing agency Randstad, only 13 percent of employees have received AI training in the past year.[7] This lack of training can lead to a lack of understanding of the critical role humans play in AI oversight. As Rosalia Tungaraza, assistant vice president of artificial intelligence at Baptist Health, aptly puts it, "If you put an AI in there, and it works five times in a row, I can see it's like: if you don't double-check it, what's the worst that can happen?"[8]

While researching for this book, I discovered references given by generative AI tools sometimes didn't exist. Even if they did exist, they didn't always support the claims cited.

To mitigate these risks, companies are developing new protocols and urging AI tool providers to include double-checking mechanisms in their workflow. The aim is to prevent the potentially disastrous consequences of unchecked AI, such as legal issues arising from incorrect company documents or customer-facing material.

The introduction of GPT-3 in 2020 unleashed a new era of natural language processing capabilities. Chatbots like ChatGPT showcase how far we've come in conversational AI. Yet for all the hype, we are still in the early innings of this technology's development and integration into business operations. Much work remains to enhance accuracy, tweak ethical parameters, and apply generative models to more specialized domains.

SECTION II

THE INTEGRATION

CHAPTER 5

A Strategic Framework for Enterprise AI Adoption and Integration

---

"The challenge of integrating AI is not a technology problem. It is a broader change management problem that involves people, processes, data and models." —Srivatsan Laxman, Senior Vice President of Global Customer Transformation at Unilever

It was a crisp fall day in Los Altos. I was sitting with John for lunch at the Holder's Country Inn. John, a tech industry veteran and the board chairman of a company, seemed preoccupied as we caught up over iced teas and Caesar salads.

"Every analyst call, every investor meeting, it's the same question—'What is your AI strategy?'" John shook his head "The company needs to have a concrete plan for integrating artificial intelligence into our product roadmap and operations." I nodded knowingly. "We've dedicated half a day at our upcoming board meeting for our CTO to discuss a plan for AI," John said. "But to be honest, I'm not sure where we should even begin."

Where to begin—that was the million-dollar question. As AI capabilities advanced at an unprecedented pace, the playbooks for successful enterprise adoption had not yet been written. I assured John that his company was not alone in grappling with this challenge.

From my vantage point as an AI adviser to dozens of organizations, I had seen firsthand the obstacles that arose when companies attempted to retrofit AI systems into legacy environments piecemeal. Successful AI transformation required a cohesive, holistic strategy anchored in a clear understanding of where and how AI could drive business value.

Over my salad, I outlined a high-level framework.

A PRACTICAL SIX-STEP FRAMEWORK

This is a practical six-step framework to harness the power of AI. It will help leaders to determine which business functions and aspects of business will benefit from AI and how to go from a simple PoC to deployment of multiple production level applications.

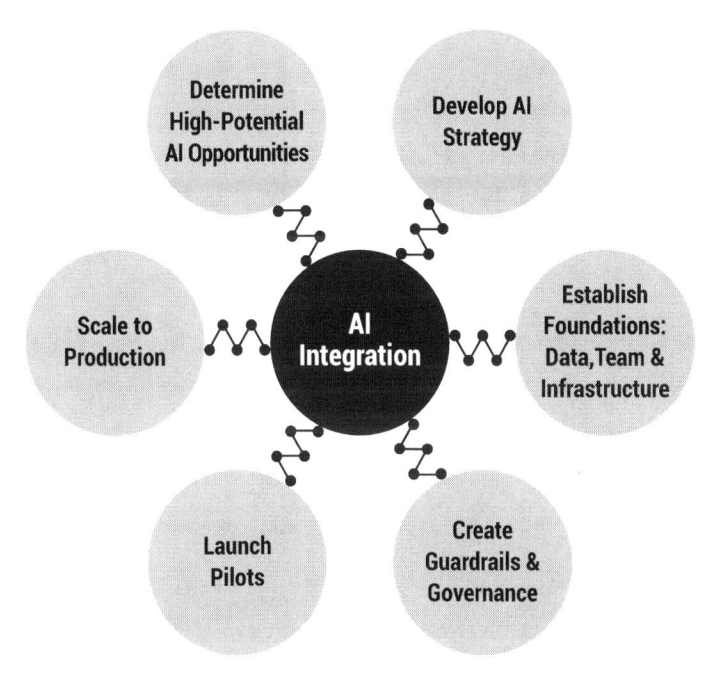

Fig 5: Six-step Framework for AI Integration

With the right strategic approach, AI can drive both top-line growth and bottom-line profitability with massive improvements in efficiency, decision-making, innovation, and customer experience and much more. The framework and advice outlined here will prepare an organization to lead the industry with the power of artificial intelligence. Let's get started!

STEP 1: DETERMINE HIGH-POTENTIAL AI OPPORTUNITIES

The starting point for an effective AI strategy is objectively evaluating potential applications across your business. By methodically scanning for opportunities, you can focus AI investments on use cases likely to drive maximum value. Here are best practices for this assessment:

CONDUCT AN AI OPPORTUNITY SCAN

Conducting an AI Opportunity Scan involves several crucial steps aimed at identifying areas within a business where AI can be effectively implemented to enhance efficiency and innovation. The process includes:

- **Forming a Cross-functional Team:** Assemble a diverse team of eight to ten members from various departments such as IT, operations, product development, customer success, marketing, and finance. This diversity ensures a comprehensive understanding of the potential AI applications across different business units.
- **Engaging Business Units:** Organize workshops with representatives from each business unit. The goal is to delve into their specific challenges, pain points, and areas ripe for improvement. This collaborative approach fosters a deeper understanding of where AI can add value.
- **Applying Analytical Tools:** Use tools like value stream mapping to dissect current operational processes. This helps in pinpointing steps that are potential candidates for automation, thereby enhancing efficiency.
- **Maintaining an Open Mindset:** Create a culture of openness and creativity among team members. This

attitude is vital for uncovering nonobvious applications of AI that might otherwise be overlooked.
- **Researching and Brainstorming:** Dive into existing AI use cases within your industry and beyond. Investigate how leading companies are leveraging AI for a competitive edge. Use this research as a springboard for brainstorming innovative AI applications that could be adopted in your organization.
- **Focusing on Key Areas:** Direct your efforts toward identifying opportunities to improve critical business aspects such as efficiency, decision-making, predictive analytics, and personalization. Explore ways to automate mundane tasks, enhance sales with predictive lead scoring, refine demand forecasting, provide actionable insights for customer success managers, and personalize the e-commerce experience.
- **Imagining New Offerings and Value Propositions:** Envision new AI-driven products, services, or customer experiences that could differentiate your business. This step is crucial for staying ahead of market trends and delivering unique value to your customers.
- **Categorizing Opportunities:** Organize identified AI opportunities into functional areas. This categorization helps in systematically approaching the integration of AI solutions across the entire organization.

This structured approach to conducting an AI opportunity scan is designed to methodically uncover areas where AI can significantly contribute to the organization's growth and efficiency. By following these steps, businesses can ensure a comprehensive evaluation of AI's potential impact, leading to informed decision-making and strategic implementation.

Business Area	AI Opportunities	Description
Customer Experience	Personalize Ecommerce	Tailor online shopping experiences based on user behavior and preferences.
	Insights for Customer Success Managers	Use AI to analyze customer data and provide actionable insights.
Operations	Automate Repetitive Tasks	Use AI to handle routine tasks, freeing up human resources for other work.
	Forecast Demand	Use AI algorithms to predict customer demand for products or services.
Product Development	AI-driven Product Design	Use AI to analyze market trends and customer feedback to design new products.
Finance	Budget Forecasting	Use AI to predict future budget needs based on historical data.
Sales & Marketing	Customer Segmentation	Use AI to segment customers into different groups for targeted marketing.
Human Resources	Talent Acquisition	Use AI to scan resumes and match candidates with job requirements.

Table 3: Opportunity Categorization

PRIORITIZE OPPORTUNITIES WITH THE LARGEST POTENTIAL BUSINESS IMPACT

When prioritizing opportunities, you must meticulously assess each prospective AI application. Your evaluation should focus on the expected value creation, the feasibility of implementation, and the alignment with your overall business strategy. For every opportunity, try to gauge its potential in generating revenue, cutting costs, boosting customer satisfaction, and delivering other significant benefits. Consider the data, technology, and capabilities your team needs to succeed.

Moreover, you should organize these opportunities into distinct phases or "waves." You'll base this organization on factors like the AI application's complexity, the expected timeline for rolling out, and the resources it will demand. This method allows you to build a structured roadmap for integrating AI into your business processes.

Initially, concentrate your efforts on pilot projects with a high likelihood of success and significant impact. These high-confidence use cases can prove the value of your AI investments and lay the groundwork for more ambitious applications in later phases. This strategic approach to implementing AI technologies maximizes the potential for transforming your business operations.

SCORE CARD SYSTEM

The score card system is a strategic framework designed to prioritize AI initiatives based on their potential benefits and feasibility. It involves a comprehensive evaluation across various criteria, each assessed on a scale from one to ten, where ten signifies the highest potential or requirement level. This evaluation aids in categorizing AI opportunities into distinct waves, reflecting the project's complexity, required timeline, and resource demands. The primary focus is on identifying and advancing initial pilots that exhibit high confidence in achieving substantial benefits upon successful implementation.

EVALUATION CRITERIA

- **Revenue Gains:** The extent to which the AI initiative can enhance revenue streams through various means such as unlocking new opportunities or changing business models.
- **Cost Reduction:** The degree to which the initiative can decrease operational costs, including labor, materials, and overheads, contributing to a leaner and more cost-effective operation.
- **Customer Satisfaction:** The impact of the AI initiative on enhancing customer experience, satisfaction, and loyalty, which is crucial for long-term business success.
- **Other Benefits:** This includes all additional advantages the initiative may bring, such as competitive edge, market position, and operational excellence.
- **Data:** The availability, quality, and integration complexity of the data necessary for the AI initiative.
- Technology: The sophistication, availability, and integration challenges of the technology needed for the initiative.
- **Capabilities:** The level of expertise and skills—including AI, data science, and change management, and resources—required to successfully execute the initiative.
- **Other Feasibility Factors:** Additional critical aspects influencing the success of AI initiatives, including organizational readiness, regulatory and ethical compliance, scalability, integration with existing systems, and risk management.

Criteria	Score Range	Waves (Complexity, Timeline, Resource Requirements)	Focus for Initial Pilots
Potential for Revenue Gains	1-10	Wave 1, Wave 2, Wave 3	High-Confidence Use Cases
Potential for Cost Reduction	1-10	Wave 1, Wave 2, Wave 3	High-Confidence Use Cases
Potential for Customer Satisfaction	1-10	Wave 1, Wave 2, Wave 3	High-Confidence Use Cases
Other Benefits	1-10	Wave 1, Wave 2, Wave 3	High-Confidence Use Cases
Required Data	1-10	Wave 1, Wave 2, Wave 3	High-Confidence Use Cases
Required Technology	1-10	Wave 1, Wave 2, Wave 3	High-Confidence Use Cases
Required Capabilities	1-10	Wave 1, Wave 2, Wave 3	High-Confidence Use Cases
Other Feasibility Factors	1-10	Wave 1, Wave 2, Wave 3	High-Confidence Use Cases

Table 4: Scorecard system

EXAMPLE: INVOICE AUTOMATION

Now let's walk through a real AI project to illustrate the scorecard system. A multinational hardware company serves over a thousand customers. Their invoice processing is very manual, and over ninety analysts process them. They involve custom processes for each client. More than 10 percent of the invoices are returned due to errors requiring reprocessing and starting the payment cycle all over again. Their payment term is sixty days, but given the processing time, payments are received typically in ninety to one hundred and twenty days. On top of that, given the manual processing, insights from the customer payments are not readily available. Potential cost savings exceed fifteen million dollars annually, errors and rework are minimized, payments are received closer to the intended timeframe, and valuable data becomes readily available for better decision-making.

SCORE CARD

- **Revenue Gains:** Accelerating the payment cycle to sixty days could indirectly boost revenue by improving cash flow and capital efficiency. Score: 7
- **Cost Reduction:** By automating invoice processing, the company can reduce manual labor and associated costs. Score: 9
- **Customer Satisfaction:** Streamlining the invoice processing improves customer experience while reducing errors and delays. Score: 8
- **Other Benefits:** Valuable data insights for better decision-making and minimize errors and rework, contributing to operational excellence. Score: 8

- **Data:** Data already exists, but integrating and cleaning this data for AI use might present challenges, especially with custom processes for each client. Score: 6
- **Technology:** The complexity of custom client processes and the need for a robust AI system to handle variability might require potentially costly solutions. Score: 5
- **Capabilities:** Needs significant AI and data science expertise to develop and implement this solution as well as change management to transition to automated processes. Score: 5. However, if the implementation is given to a third party, the score could be 8.
- **Other Feasibility Factors:** Scalability is a challenge for this project due to the complexity of rules customized for each customer. This could be divided into phased implementation. Score: 4

While there are significant cost, customer experience, and other benefits, the complexity associated with data, technology, capabilities, and scalability doesn't make it an easy project to implement. Overall, this is a Wave 2 project.

Criteria	Score	Wave	Focus for Initial Pilots
Potential for Revenue Gains	7	Wave 2	No
Potential for Cost Reduction	9	Wave 1	Yes
Potential for Customer Satisfaction	8	Wave 1	Yes
Other Benefits	8	Wave 1	Yes
Required Data	6	Wave 2	No
Required Technology	5	Wave 2	No
Required Capabilities	5 or 8	Wave 2	Maybe
Other Feasibility Factors	4	Wave 2	Maybe

Table 5: A Scorecard Example

STEP 2: DEVELOP AN AI STRATEGY

Treating AI as the domain of IT or a CTO is shortsighted. So is treating AI as a vehicle to cut costs. Properly deployed, AI enables new business models and opens avenues for growth including new markets. It also brings operational efficiencies and unprecedented personalization while creating outstanding customer experience.

Given these enormous opportunities to add to growth and profitability, AI should be part of a company's long-term strategy. Think boldly! Why not aim for a customer satisfaction score of 95 percent? Why not aim for a productivity gain of 70 percent or instantaneous closing of books on the next day of the quarter close? Of course, enabling AI inside a company takes experimentation, failures, successes, and adjustments from lessons learned. It will be a multiyear journey.

ALIGN AI PROJECTS TO BUSINESS GOALS AND STRATEGIES

A robust AI strategy is crucial to ensure initiatives are in harmony with broader business objectives, efficiently utilize resources, and set a path for expanding successes. It's essential to prioritize solving real business issues over the allure of technology itself, ensuring outcomes contribute to the overarching business aims, regardless of the project's scale. A comprehensive AI strategy should encompass various facets, including business, organizational, infrastructure, technology, and governance while contemplating both immediate and future sourcing strategies.

THE PITFALLS OF LACK OF STRATEGIC ALIGNMENT

The absence of long-term planning and a strategic approach is a significant reason many companies stumble in their AI adoption journey. Companies often rush into AI projects without considering how they align with their core competencies or further critical business goals. This lack of strategic alignment means AI initiatives often fail to solve real business problems, leading to wasted time, effort, and financial resources. According to the Project Management Institute (PMI), 44 percent of projects fail due to a lack of alignment with business goals.[1]

For example, consider a hypothetical consumer electronics company that decides to leverage AI for predictive analytics. Despite the technical robustness of their models, they lack a clear strategy for how these analytics will improve supply chain efficiency. The consequence? The AI initiatives are discarded after the pilot phase, squandering the data science team's time and resources. Consider another scenario where a hypothetical retail company adopts AI to offer personalized product recommendations to boost sales. However, the AI team did not collaborate effectively with store managers and marketing teams to grasp the actual needs of the customers. As a result, the AI system ends up recommending products that don't align with customer preferences, leading to dissatisfaction instead of boosting sales.

ENSURING STRATEGIC ALIGNMENT

To align AI initiatives with broader corporate aims, review the overarching strategies and primary goals for the coming three to five years, pinpointing AI opportunities that directly bolster these objectives. Demonstrating how AI endeavors will meet key targets is essential. Engage with leaders from different business units to grasp their urgent needs and challenges, integrating this insight into the AI strategy and planning. Establish three to five AI-specific strategic goals for your organization. For instance, a logistics company might use AI to optimize delivery routes based on traffic conditions, weather forecasts, and package delivery priorities.

This not only improves delivery efficiency but also reduces fuel consumption and operational costs, aligning with the company's goals of increasing operational efficiency and sustainability.

Companies should resist the temptation to chase after AI simply because it's the latest buzzword or because vendors are pushing the newest algorithms. Instead, the focus should be on how AI can add tangible value to the business. To ensure AI projects are strategically aligned, organizations should adopt specific, measurable, achievable, relevant, and time-bound (SMART) goals. For example, rather than vaguely aiming to "improve customer service," a more focused SMART goal would be "to reduce customer service response times by 30 percent within six months using AI-powered solutions." This not only provides a clear direction but also sets the stage for measurable outcomes and accountability.

SECURE EXECUTIVE BUY-IN

To maximize AI's value, it's essential to identify and quantify the top AI opportunities, focusing on their impact on competitiveness, customer experience, financial, and strategic perspectives. Compiling these insights into a concise report for executive leaders is crucial, emphasizing how AI initiatives align with the company's strategic goals. Securing executive support for two to three priority AI projects is crucial for their successful integration. Conducting an objective evaluation of AI opportunities builds confidence in these investments and ensures the allocation of resources to the most impactful areas first, driving meaningful organizational advancement.

BUILD MULTIYEAR AI PROJECT ROADMAP

Sequence the high-value AI opportunities identified in Step 1 into a realistic implementation roadmap over the next three to five years. Group AI projects into successive waves, with each wave building on capabilities and lessons learned from previous ones. Define target milestones for piloting, production deployment, and measuring business impact for each wave. Factor in dependencies across projects. Leave room to adapt based on pilot results.

STEP 3: BUILD FOUNDATIONS: DATA, INFRASTRUCTURE AND TEAM

While piloting focused applications, it's also important to start establishing the data, security, and infrastructure foundations required to scale AI across the enterprise.

The purpose of establishing these foundations is to ensure that AI projects are built on a solid base of high-quality data, supported by the necessary technological infrastructure, and driven by a skilled team. The benefits include enhanced decision-making, improved operational efficiency, and the ability to innovate and maintain a competitive edge through the strategic use of AI. By focusing on these foundational elements, organizations can better position their AI initiatives for success, enabling scalable and sustainable integration of AI technologies across their operations.

ASSESS AND IMPROVE QUALITY OF DATA FOR MODEL TRAINING

To assess and improve the quality of data for model training, the approach starts with cataloging existing data sources and formats within the company. This step is crucial to understand the landscape of available information and its current organization. Identifying gaps for AI use cases involves pinpointing where the available data does not meet the requirements or where additional data could enhance AI applications, guiding targeted data collection or generation efforts.

Analyzing data health—including completeness, accuracy, and bias checks—ensures the data used for training AI models is reliable and representative. This is critical because the quality of input data directly impacts the performance and fairness of AI systems. Cleansing and enriching data as needed involves removing errors, inconsistencies, and irrelevant information as well as augmenting the dataset with additional relevant data to improve its quality and usefulness for AI applications.

Planning data collection processes to fill gaps through methods like surveys, IoT sensors, or synthetic data is about strategically increasing the dataset's volume and variety to cover all necessary aspects for the AI's intended function. This step ensures that the AI system can perform well across a wide range of scenarios and inputs.

Designing data pipelines to consolidate, prepare, label, and serve data for model development and production deployment is about creating efficient, automated processes that handle data from collection through usage in AI models. This includes preprocessing steps, like normalization and feature extraction, as well as ensuring data is in a usable format for training AI models.

Implementing data lifecycle governance—including security, privacy, lineage, and retention policies—ensures data handling complies with legal and ethical standards throughout its lifecycle, from collection to disposal. This governance is essential for maintaining trust in AI systems and ensuring they are used responsibly.

These steps collectively ensure the data used for training AI models is of high quality, leading to improved model performance, reduced biases, enhanced decision-making, and compliance with regulatory and ethical standards. More details can be found in chapter 6 "Fuel for Thought: The Cornerstones and Challenges of Data in AI."

EVALUATE COMPUTING INFRASTRUCTURE FOR AI NEEDS

Take inventory of existing on-premise servers and cloud platforms currently used for analytics and applications. Analyze technical specifications of current infrastructure like CPU, GPU, memory, and storage. Run representative AI workloads at small scale to test performance and limitations of current infrastructure. Estimate order of magnitude, additional processing power, faster GPUs, and scalable storage needed for enterprise-wide AI based on projections.

Provision additional GPU/TPU processing, scalable storage, ML Ops software as required. Upgrade on-premise data center servers optimized for parallel processing AI/ML workloads. Leverage cloud provider services like AWS, Azure, and Google for dramatically faster training and inference. Allocate additional storage capacity on-premises and in cloud to support versioned ML model repositories, training datasets, and model outputs. Acquire MLOps software platforms, such as Algorithmia or Spell, to operationalize model monitoring, retraining, tuning, and deployment.

Integrating technologies like machine learning (ML), natural language processing (NLP), and automated machine learning (AutoML) into an enterprise AI framework is crucial for several reasons. ML allows for the analysis and interpretation of complex data patterns, enhancing decision-making and operational efficiency. NLP enables the processing and understanding of human language, improving customer interactions and service automation. AutoML simplifies the model-building process, making AI accessible to nonexperts and speeding up the development of AI solutions.

The integration of these technologies fosters a versatile AI ecosystem that can tackle a broad range of business challenges—from customer service to predictive analytics. This ensures enterprises can leverage AI not only to optimize existing processes but also to innovate and create new business models while staying competitive.

Perform external cybersecurity audits to identify vulnerabilities in existing IT infrastructure. Harden networks, implement zero-trust access controls, and encrypt data to protect IP and customer data. Develop model governance policies for secure data sourcing, labeling, and compliant model use. Build tools to monitor data and model use, log access, detect anomalies and trigger alerts. Institute disaster recovery provisions for resilient AI system operations. Comprehensive coverage of security is covered in chapter 7 "An Ounce of Prevention: Securing the Competitive Edge of AI."

BUILD OR SOURCE

Deciding whether to build AI capabilities in-house or source them from external providers involves evaluating factors such as cost, expertise, and strategic importance. For core competencies, building in-house may offer greater control and alignment with business goals. Conversely, sourcing can expedite deployment and access to cutting-edge technologies, especially in noncore areas.

ESTABLISH THE RIGHT AI TEAM

Designate or hire a senior executive like a chief AI officer to drive AI strategy across the company. The chief AI officer should be a business and technology expert with a strong understanding of how AI can be leveraged to create business value and the technical aspects of AI, rather than someone who is solely focused on creating new AI models or algorithms. Establish a guiding coalition of AI leaders from business units, IT, operations, finance, risk management, and other groups. Ensure alignment and shared commitment. Empower the team to remove blockers, secure resources, and work cross-functionally to achieve AI success. Determine the skills gap for building, deploying, and maintaining AI models including data scientists, ML engineers, model trainers, etc.

Partner with HR to develop a hiring plan including sourcing, candidates screening, and competency assessment. Consider contracting specialists like data annotation teams to accelerate progress. Conduct skills assessment to identify employee groups needing AI and data training. Develop curriculum topics like AI fundamentals, ethics, and analyzing model outputs. Deliver training through online courses, instructor-led workshops, hackathons and other formats. See more on this topic in chapter 8 "The Art of AI Leadership: Building AI Teams."

STEP 4: CREATE GUARDRAILS AND GOVERNANCE

Institute model validation, monitoring, transparency, ethics review, and other AI governance processes proactively. Create frameworks for transitioning successful pilots into full-scale production and enterprise deployment. Plan for user training and support. Enable continuous model improvement through new data, monitoring failures, assessing biases, and regular retraining or tuning.

Rigorous model governance and controls are necessary to ensure ethical, fair, compliant, and responsible use of AI. Well-defined policies and standards mitigate risks like bias while building trust. Appoint a head of AI ethics, create model risk frameworks, and document control requirements following regulations like GDPR.

Conduct ethics training for AI teams and implement bias testing tools. Catalog approved datasets. Require data reviews before use in training. Build pipelines to scrub, anonymize, and fix missing fields. Compliance check outputs before release. Establish MLOps change control processes for reviewing model updates. Simulate edge cases using techniques like Monte Carlo to detect harmful biases. Maintain human monitoring and intervention capabilities. See chapter 9 "AI for Good: Developing AI Responsibly and Ethically" for more details.

STEP 5: LAUNCH TARGETED AI PILOT PROJECTS

With a strategy and roadmap in place, it's time to begin launching targeted AI pilots. The mantra should be: "Think big, start small, scale fast." Pilots allow you to demonstrate value and build organizational capabilities in a focused manner before approving full-scale production. A well-executed pilot helps strengthen executive sponsorship, build employee enthusiasm, and avoid overcomplicating solutions early on.

BENEFITS OF STARTING SMALL

Starting with small, tightly scoped pilots has several key advantages over attempting large-scale AI overhauls from the outset. Small projects derisk initiatives by allowing experimentation in a low-risk environment without committing extensive resources upfront. They provide a testing ground to validate hypotheses, run trial algorithms, identify potential challenges, and gather critical insights to inform future scale-up efforts.

For example, last year my team identified over twenty potential AI use cases for a global hardware company. We chose automating contract obligations using generative AI as the initial proof-of-concept project. This four-month pilot is projected to deliver millions in annual savings. More importantly, it surfaced invaluable experience, navigating process, infrastructure, change management, and system integration hurdles—lessons that will streamline subsequent AI projects.

START WITH TIGHTLY SCOPED PILOTS FOCUSED ON HIGH-VALUE OPPORTUNITIES

Maintain discipline to pursue narrow, well-defined pilots before expanding scope. Pilots with too broad a scope often fail. Focus initial pilots on high-confidence AI opportunities expected to have sizable business impact based on ROI assessment. These are the Wave 1 projects identified earlier.

Take advantage of pre-built machine learning models and AI services through third-party providers to accelerate development. Integrate pre-built solutions as feasible versus building custom models from scratch. Third-party AI software space is evolving rapidly and make sure to check established as well as smaller companies that may have custom solutions, which may fit your needs more closely. Make sure to understand the risk tolerance of your company in using outside solutions and the risk of using solutions of smaller but highly innovative companies.

Assemble small pilot teams with a combination of business, technical, design, and other skill sets. Adopt agile principles like sprints, regular demos, and continuous user feedback to drive rapid iterations. Empower teams to make decisions independently and move fast.

DEVELOP METHODS TO MEASURE PILOT SUCCESS AND LESSONS

Define clear success metrics upfront aligned to specific business metrics like improving customer satisfaction scores by 5 percent or achieving 90 percent algorithm prediction accuracy on test data. Design evaluation mechanisms to capture successes and key lessons while addressing technical debt and user pain points.

Create playbooks documenting processes followed, obstacles faced, pivotal decisions made, and recommendations given to inform future pilots. Be prepared to adjust or stop pilots not meeting criteria to refocus resources. These focused pilots establish critical organizational AI capabilities and a path to production.

STEP 6: SCALE ACROSS THE COMPANY TO MAXIMIZE BUSINESS BENEFITS

Transitioning pilots to production is important to achieve the full business benefits validated during the pilot phase. AI capabilities remain siloed without concerted effort to integrate models into workflows and scale impact across the organization. Before taking projects to production, it's important to have the scalable data and tech infrastructure and AI team in place.

Perform code reviews of pilot systems to identify optimizations needed for scale. Refactor models for improved performance based on lessons learned. Build in monitoring, logging, and alarms using DevOps tools such as Jenkins, Grafana, and ELK. Embed models into apps, workflows, and processes through APIs, microservices, and other integration patterns. Present outputs through dashboards, notifications, and UIs tailored to user context. Conduct usability testing to refine.

Create dedicated AI Operations teams with agile practices like daily standups, retrospectives, and backlog grooming. Develop runbooks and on-call schedules for twenty-four-seven model monitoring and support. Implement gradual rollouts to progressively ramp production.

IMPLEMENT ONGOING MODEL MONITORING AND MAINTENANCE

Continuous model monitoring and maintenance of AI models is required to sustain accuracy and value delivery over time. Models decay without diligent governance to refresh training data, tune hyperparameters, and upgrade algorithms. Build regression testing suites to continually validate model accuracy. Set performance thresholds and alerts in tools like DataRobot and TruEra.

Schedule model tuning sprints every quarter. Budget time for experimenting with state-of-art techniques and frameworks like TensorFlow and PyTorch. Implement continuous integration/continuous delivery (CI/CD) pipelines to retrain models on new data. Analyze trends in prediction quality and flag inputs causing errors to the ML engineering team. Refresh models incrementally on daily data versus rebuilding annually. Use A/B tests to validate improvements.

KEEP SOLUTIONS SIMPLE AND EFFECTIVE

Early in my career, a colleague's advice at Intel stuck with me: "Keep it Simple and Stupid." It's a reminder not to overcomplicate solutions unnecessarily, especially in the tech world where we often face a temptation to build the most advanced, intricate solutions possible.

However, complexity increases the chances of errors and makes systems harder to understand, maintain, and update. A simple, well-designed solution can often deliver the same results as a complex one but with fewer headaches.

To avoid overcomplication, adopt a user-centric design approach, understanding end-users' needs and limitations. An intuitive and easy-to-use system is more likely to be adopted successfully.

To embrace an iterative development approach, start with a basic version, test it in real-world conditions, and refine it based on feedback. This ensures simplicity and alignment with actual business needs.

Remember, complexity isn't always better. Keep solutions simple and effective.

SHARE BEST PRACTICES TO DRIVE SCALE

Propagating know-how and assets from early AI projects accelerates adoption across the company. Teams shouldn't reinvent the wheel when proven playbooks and templates are available. Identify AI program alumni to consult with teams new to AI. Allow them to function as coaches to provide hands-on guidance and architectural reviews. Curate gallery of demos to showcase successful use cases and the art of possibility with AI. Record testimonials on how AI increased sales and cut costs to prove value. Develop and share templates for model cards, ML requirements, data reports, and testing plans to propagate reusable components.

SCALING WITH AI OPS

Incorporating AIOps (artificial intelligence for IT Operations) practices can effectively manage and scale AI projects. AIOps leverages machine learning to automate IT tasks, analyze vast data to identify patterns and predict issues, and streamline incident response. Benefits include:

- Automating infrastructure monitoring and management, freeing IT resources for higher-level tasks, detecting anomalies, and predicting potential issues.
- Improving data quality and observability by continuously monitoring data pipelines for errors and biases, ensuring quality data for AI models and facilitating better understanding of model behavior and root cause analysis.
- Optimizing resource allocation by analyzing utilization and identifying optimization opportunities, ensuring optimal performance while reducing unnecessary costs.

DRIVE CONTINUOUS IMPROVEMENT

Sustaining progress requires continually identifying fresh AI opportunities while scaling successes. New use cases and technologies keep capabilities advancing across the organization. Conduct quarterly AI opportunity workshops enabled by AI Centers of Excellence (CoE). Maintain idea funnel using AI opportunity assessment process. Incentivize participation through rewards and recognition of teams adopting AI. Prioritize new pilots balancing business value, speed to deploy, and diversity of use cases. Sunset obsolete models gracefully while ramping up new releases.

Transforming your business with AI demands careful planning, disciplined implementation, and ongoing dedication. The six-step framework provided—from identifying valuable AI opportunities and developing a strategic roadmap to initiating pilot projects, establishing foundational capabilities, and broadening successful initiatives organization-wide—facilitates meaningful AI integration for tangible business benefits. This journey, while demanding patience and continuous commitment, offers substantial rewards for organizations ready to embrace it, enabling them to enhance efficiency, decision-making, innovation, and customer experiences as well as to ascend to new performance peaks with AI as a pivotal lever in their strategic arsenal.

AI integration is filled with potential stumbling blocks, but awareness and preparation can help to avoid them. With strategic alignment, high-quality data, robust training cycles, cross-functional teams, and ongoing monitoring, companies can feel confident in unlocking AI's possibilities. Of course, the integration journey requires flexibility and course corrections.

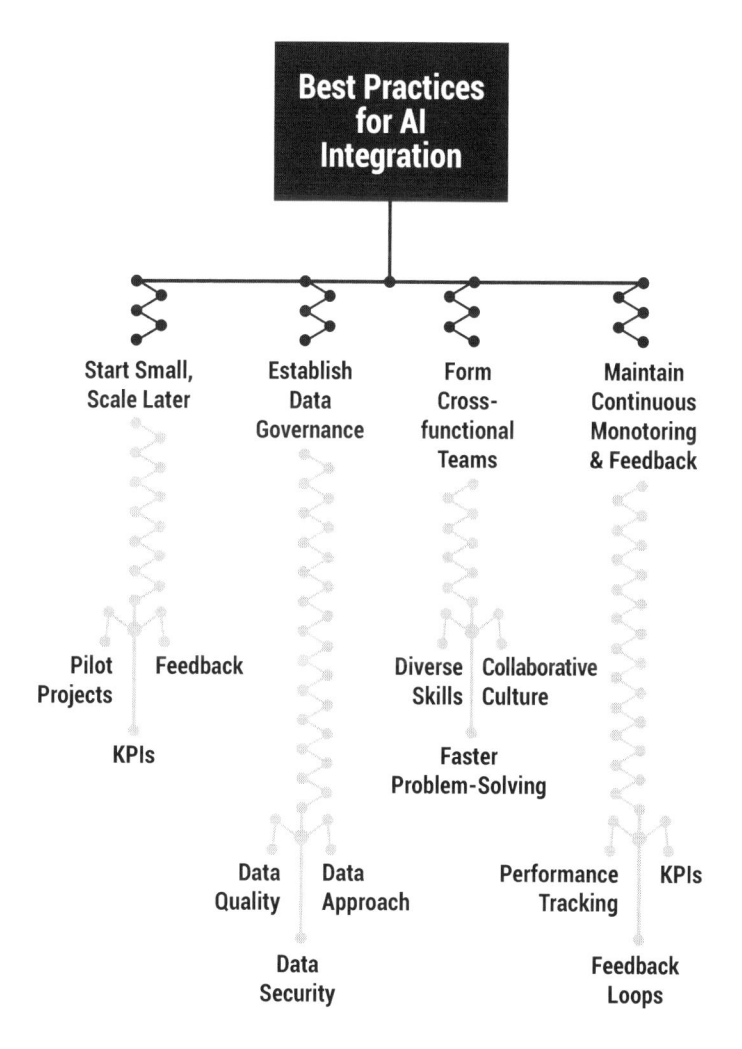

Fig 6: Best Practices for AI Integration

CHAPTER 6
Fuel for Thought: The Cornerstones and Challenges of Data in AI

———

In early 2020, James Collins and Regina Barzilay were leading a research team at MIT and made a groundbreaking discovery—an antibiotic capable of killing strains of bacteria that were resistant to all other known antibiotics.[1] The collaboration of human expertise and artificial intelligence powered this extraordinary feat. Traditionally, the discovery of a new antibiotic is a laborious and time-consuming process. Researchers start with thousands of possible molecules and, through a series of trial and error, narrow them down to a few viable candidates. This method is not only expensive but also highly uncertain as it relies heavily on educated guesses and serendipity.

The researchers employed deep learning, and data played a pivotal role in this discovery. The team gave the algorithm a vast amount of information, including the molecular structures of existing antibiotics, data on bacterial resistance to existing drugs, and the chemical properties of various compounds. The first step in the process was to train the model using a labeled dataset. This dataset contained molecular structures of known antibiotics and other compounds, each tagged with its effectiveness against bacteria. They integrated the data from disparate sources into a unified database. This allowed the AI algorithm to have a 360-degree view of the problem space. The AI algorithm used this rich dataset to make informed predictions. This is an example of supervised learning we discussed in chapter 3. The algorithm learned to associate specific molecular features with antibacterial efficacy. For example, it could identify a molecule developed for an entirely different purpose, like diabetes treatment, and recognize its potential as an antibiotic.[2]

After the training phase, the model was validated using a separate dataset not seen during training. This step ensured the algorithm could generalize its learning to new, unseen data. This comprehensive dataset enabled the AI to make highly informed decisions, drastically reducing the number of molecules that needed to be tested in the lab. The AI identified a molecule named halicin, which was initially developed as a potential treatment for diabetes. Lab tests confirmed that halicin was highly effective in killing antibiotic-resistant strains of bacteria. The discovery was groundbreaking. James Collins, one of the study's senior authors, said, "We wanted to develop a platform that would allow us to harness the power of artificial intelligence to usher in a new age of antibiotic drug discovery."[3]

THE CENTRALITY OF DATA IN AI: A CLOSER LOOK

As the halicin example illustrates, AI learns from data. If you're teaching a child to recognize fruits, you'd likely start by showing them a variety of fruits—apples, bananas, oranges—and each time you'd say, "This is an apple," or "This is a banana." Through this repetitive process, the child learns to identify each fruit. The more quality data you feed into an AI system, the better it becomes at understanding, predicting, and making decisions. In essence, data serves as both the textbooks and the real-world experiences that shape AI's "knowledge."

Why is data so crucial? Let's consider Netflix as a real-world example. The streaming giant uses data to understand what shows and movies are popular among different demographics and why. Is it suspense, the characters, or the storytelling? By analyzing this data, Netflix can make more informed decisions about what kind of content to produce or license—essentially reading the room on a massive scale.

STRUCTURED VERSUS UNSTRUCTURED DATA: THE BUILDING BLOCKS OF AI'S MASTERY

Data comes in many shapes and sizes, and we can broadly categorize it into two types: structured and unstructured. One of the most remarkable capabilities of artificial intelligence is its ability to process and analyze both, making AI an invaluable tool for businesses.

Structured data is highly organized and easily searchable, much like the books in a library catalog. This type of data is often found in databases where customer information is stored or in spreadsheets that detail a company's monthly expenses. It's a powerhouse for predictive analytics. AI algorithms can sift through structured data to identify patterns and make accurate forecasts. Take the stock market, for instance. The data here is highly structured with specifics like opening prices, closing prices, trading volumes, and more. AI can analyze years of this data to predict future market trends, giving investment firms a competitive edge.

On the flip side, unstructured data is more free-form and doesn't fit neatly into tables or spreadsheets—a content jungle of things like emails, social media posts, videos and more. While it may not be organized, its value lies in the qualitative insights it offers for understanding consumer sentiment and behavior. Envision a tech company launching a new smartphone. The company would be keen to know the public's reaction, which AI can gauge by scanning through the unstructured data of social posts, reviews, comments, and more. This unstructured data analysis allows marketing and product teams to adjust strategies based on whether the public sentiment is positive or negative.

Both data types have immense value for training AI systems. Structured data provides clean fuel for building powerful predictive models. Unstructured data gives a raw, unfiltered view into real-world usage and behavior. Combined, they provide a complete 360-degree view for businesses. The foundations require cross-department collaboration, with IT overseeing structured databases while marketing or others manage unstructured content—all working together to build robust data pipelines to feed AI applications.

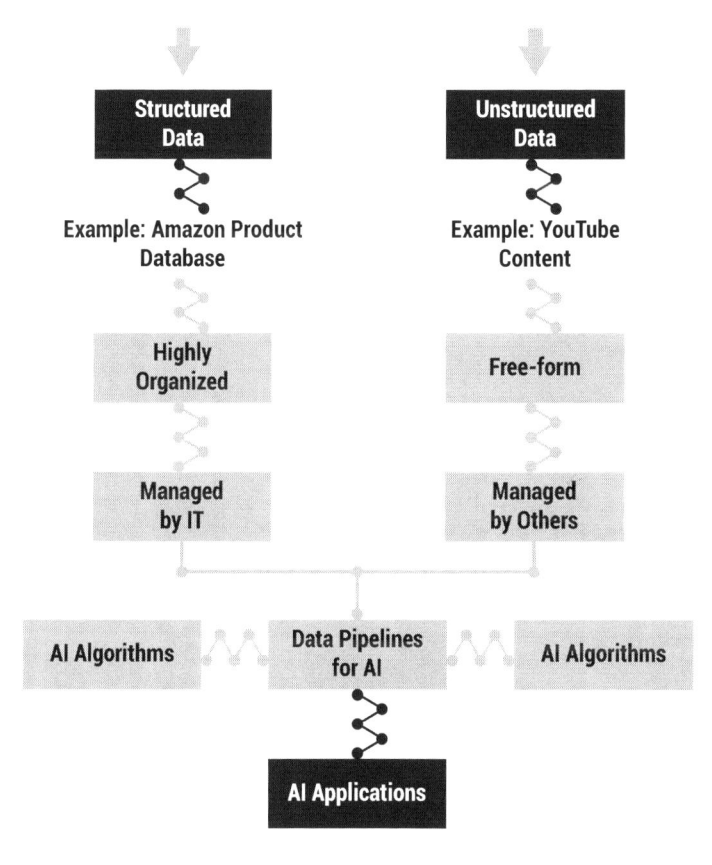

Fig 7: Structured and Unstructured Data

POTENTIAL DATA PITFALLS IN AI INTEGRATION: NAVIGATING THE MINEFIELD

While data is the lifeblood of AI, it's crucial to be aware of the potential impact of data on AI models.

DIRTY DATA: THE MOVIE RECOMMENDATION FIASCO

Recently I was catching up with Kristin, a former colleague, who told me about an AI movie chatbot a friend had raved about. Her friend loved romantic comedies and had told Kristin that the chatbot's suggestions were spot on. When Kristin logged in and listed her preferences for feel-good, lighthearted films, the AI chatbot immediately recommended the new release *Nightmare at Camp Bloodbath*. Confused, Kristin checked the movie's description. It was definitely a gory horror flick, the complete opposite of what she wanted. Apparently, some data errors had occurred, and someone mislabeled the movie's genre. So much for the credibility of the company that created the AI chatbot. She had been looking forward to a fun movie night with her husband Mike, but the bad recommendation ruined those plans.

If the data fed into this system is incorrect or "dirty," the recommendations can go horribly wrong. What does "dirty data" mean? It might sound a bit abstract, so let's break it down. In the context of AI, data can be considered "dirty" for several reasons:

Incomplete Data—Suppose you're running an e-commerce business, and your customer database is missing some values—age or location, for some users. If your AI model relies on these attributes for personalized recommendations, the incomplete data can lead to less accurate results.

Inconsistent Formatting—When you're analyzing customer reviews from multiple platforms, and if one platform rates products out of five stars and another out of ten, the inconsistency can confuse the AI model. Standardizing these ratings to a common scale is crucial for accurate analysis.

Outliers—In a dataset, an outlier is a value that is significantly different from other values. For example, if you're analyzing monthly sales data and see a sudden spike due to a one-time event like a clearance sale, that spike is considered an outlier that deviates significantly from the typical sales pattern. If not handled correctly, outliers can skew the AI model's understanding of "normal" behavior.

Duplicate Entries—Duplicate data can artificially inflate the importance of certain records. For instance, if a customer's purchase is recorded twice due to a system error, an AI model might incorrectly interpret this as two separate purchases, affecting its future recommendations.

Incorrect Labels—In supervised learning, AI models rely on labeled data to learn. If the labels are incorrect, the model will learn incorrectly. For example, if a cat is labeled as a dog in a dataset used to train a pet recognition AI, the model might incorrectly identify cats as dogs.

Human Errors—Simple human errors in data entry, like typos or incorrect units (for example, meters instead of feet), can also make data "dirty." These errors can significantly impact the performance of AI models that rely on precise measurements.

Temporal Irrelevance—Data that is outdated or not aligned with the current time frame can also be considered "dirty." For example, using fashion trends from the 1990s to predict what will be popular in 2023 would likely lead to inaccurate predictions.

Dirty Data Types	Example	Impact
Incomplete Data	Missing values for age or location	Less Accurate AI Models
Inconsistent Formatting	Ratings out of different scales	Confused AI Models
Outliers	Sudden spikes in sales data	Skewed AI Understanding
Duplicate Entries	Double recording of purchases	Inflated Importance
Incorrect Labes	Mislabeled data in supervised learning	Incorrect Leraning
Human Errors	Typographical errors	Impacted Performance
Temporal Irrelevance	Outdated fashion trends	Inaccurated Predictions

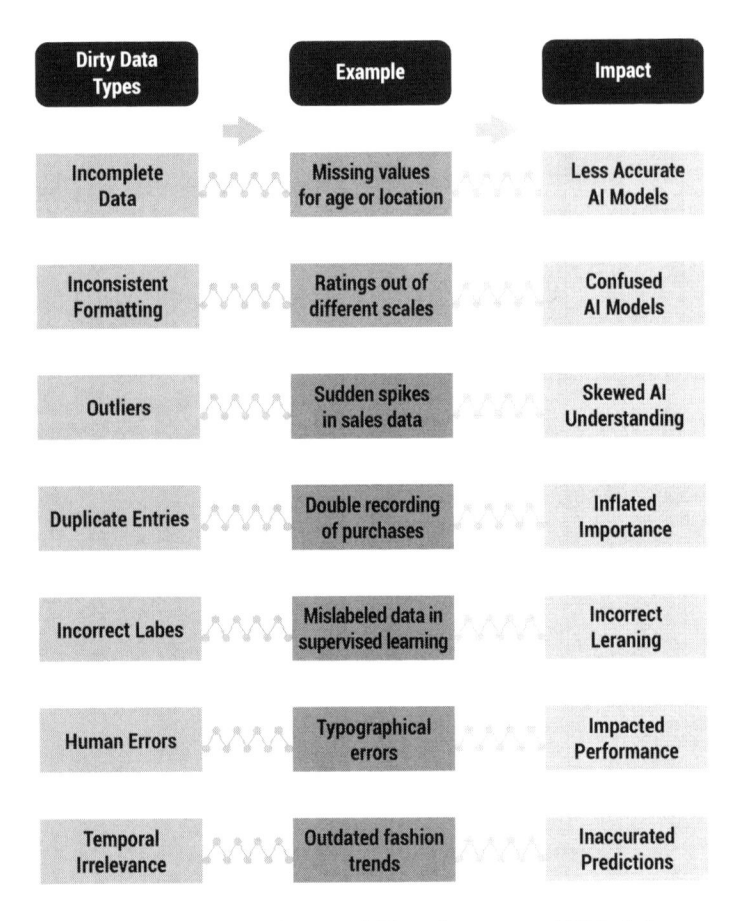

Fig 8: Dirty Data Types and Their Impact on AI Models

THE IMPORTANCE OF DATA CLEANING

THE CHALLENGE OF DATA QUALITY

Rushing into AI without proper data is a common downfall. Many AI initiatives flounder because of problematic data that leads to poor model performance. If the data is incomplete, noisy, biased, or overly complex, even the most advanced algorithms will fail. Companies often underestimate the investments needed in data infrastructure and governance required for enterprise-wide AI.

A manufacturing company wanted to implement predictive maintenance to reduce equipment downtime. However, they stored their operational data in multiple legacy systems rife with gaps, inaccurate sensor readings, and manual data entry errors. Data engineers had to spend months stitching together pipelines and cleaning data before the data scientists could build models. By then, the AI prototypes were already behind schedule and over budget.[4]

Similarly, an anonymous financial services firm that we call XYZ Inc, struggled with an AI-based fraud detection system that kept flagging legitimate transactions as fraudulent. Upon investigation, the AI team found biases in the historical transaction data used for training. Their models learned to associate certain zip codes and customer names with fraud just because past data correlated those features with fraud flags from rule-based systems.

ADDRESSING DATA PROBLEMS

Solving these data problems requires upfront investment into data integration, monitoring, governance, and the IT infrastructure needed to manage thousands of data points. Companies cannot expect to throw messy, biased data into black-box AI and achieve business value. With the right data foundation, AI can transform operations. But without it, AI will languish as an experiment. To mitigate these risks, organizations should implement robust data governance policies. This involves standardizing data formats, ensuring data accuracy, and conducting regular audits. Additionally, before deploying any AI model, it's crucial to validate the data with subject matter experts in the relevant field. For instance, in a health-care setting, medical professionals should verify the data's accuracy and relevance because if they don't, AI models could make incorrect diagnoses or treatment recommendations, potentially endangering patients' health.

By giving due attention to data quality, organizations can significantly increase the chances of AI project success. High-quality data not only improves the accuracy of AI models but also ensures they are aligned with business objectives, thereby creating real value for the organization.

Given these challenges, data cleaning is a critical step in preparing for AI model training. It involves identifying and correcting (or removing) errors and inconsistencies in data to improve its quality. By investing time and resources into cleaning your data, you're setting the stage for more accurate and reliable AI models, which is essential for making informed business decisions. Understanding what makes data "dirty" can help business leaders appreciate the importance of data hygiene, ensuring that your AI initiatives are built on a solid foundation.

A CASE STUDY: NETFLIX'S APPROACH

Netflix allocates substantial resources to data verification, understanding that the quality of data directly impacts the efficacy of their recommendation algorithms. To ensure the highest data integrity, Netflix employs a variety of techniques—from automated data validation checks to manual reviews. For instance, they have developed an extendible and customizable framework designed to host and manage automated validation checks and fixes, which can identify and help address problems within specific workflows. This framework is used to ensure successful outcomes in large-scale operations. Moreover, Netflix adopts a dynamic approach by continually refining their algorithms based on real-time user feedback and behavior. This iterative process not only improves the recommendation system but also enhances customer satisfaction and engagement.[5]

DATA BIAS IN AI: REAL-WORLD CONSEQUENCES AND SOLUTIONS

In 2014, Amazon developed an AI-powered recruitment tool aimed at automating the résumé-screening process. The company trained the AI model on résumés submitted to Amazon over a ten-year period. After the training, the AI system started to favor male candidates over female ones for technical roles. The reason? The historical data fed into the AI system was predominantly male, reflecting the male-dominated tech industry. Amazon realized that the tool was not rating candidates in a gender-neutral way. For example, résumés that included words like "women's," as in "women's chess club captain," were downgraded. Amazon eventually disbanded the AI recruitment tool in 2018. They took steps to improve the diversity of their training data and are cautious about the potential for bias in AI systems.[6]

In another example, a study published in the journal Science in 2019, revealed that a health-care algorithm widely used in the US was biased against Black patients. The algorithm assigned risk scores to patients to prioritize them for various health-care programs. It used health-care costs as a proxy for health-care needs. However, due to systemic inequalities, Black patients generally incur fewer costs than white patients with the same conditions. The bias in the algorithm meant that Black patients were less likely to be referred to programs that could provide them with extra medical care, even when they were sicker than white patients. Optum, the company that developed the algorithm, acknowledged the issue and committed to addressing the bias. "This is a systematic feature of the way this kind of software is developed," said Dr. Ziad Obermeyer, one of the authors of the study, highlighting the need for more inclusive datasets.[7]

These examples show the importance of unbiased data in AI systems. Companies can combat bias in data by using diverse training data and regularly auditing the AI system's decisions for any signs of bias.

When diverse and unbiased data is hard to come by, synthetic data offers a viable alternative. Synthetic data is artificially generated but mimics the characteristics of real-world data. It can be customized to represent different demographics, thereby reducing the risk of bias in AI models. MIT researchers developed a data synthesizer that can produce synthetic datasets, allowing for the training of more equitable machine learning models.[8]

By leveraging synthetic data, companies can not only improve the accuracy and reliability of their AI models but also ensure they are ethical and fair. This is particularly important for businesses aiming to be socially responsible while also gaining a competitive edge through AI.[9]

DATA CONSIDERATIONS FOR MODEL TRAINING
DATA OVERFITTING OR UNDERFITTING

Overfitting occurs when a model is too narrowly trained on a specific dataset, causing it to perform well on the training data but poorly on new, unseen data. Let's say an AI system is trained to predict weather based solely on data from the Sahara Desert. Its predictions would always be sunny days. But what happens when this system is used to predict weather in Seattle? The AI system would fail miserably, having been too narrowly trained on a specific dataset. This is an example of overfitting.

Conversely, underfitting happens when the AI model is too simple to capture the underlying trends in the data. A weather-prediction model trained only on temperature data may fail to accurately forecast storms or rainfall, as it lacks information on other relevant variables.

To ensure robustness and generalizability, AI models should be trained on diverse datasets that encompass a wide range of variables. For example, if developing an AI model for agricultural predictions, it should not only include data from various geographical locations and climates but also consider different soil types, crop varieties, farming practices, and seasonal variations. Additionally, data should be collected over multiple years to account for annual fluctuations in weather patterns and agricultural outputs. This approach helps prevent both overfitting—where the model performs well on its training data but poorly on unseen data—and underfitting—where the model is too simplistic to capture underlying trends. By integrating comprehensive and varied data, the model can deliver more accurate and universally applicable predictions, enhancing its utility across different environments and scenarios.

DATA DRIFT

Data drift refers to changes in data distribution patterns over time. As the real world evolves, so does the data being collected from it. These shifts can diminish the accuracy of AI systems if models are not updated accordingly. For example, think of a fashion retailer that built an AI model to predict customer purchase behavior based on data from five years ago. Back then, skinny jeans were hugely popular. The AI learned that skinny jeans led to more purchases. Fast-forward to today. Wide legs, relaxed fit pants are now trending. But the old AI model still associates skinny jeans with higher sales. This data drift would result in poor performance.

Here are more examples of data drift. Consumer sentiment about a brand could change due to PR crises or new product releases. Sentiment analysis models need retraining. Demographic data like income levels and populations shift over time. Models predicting loan default risk need to adapt. New spam tactics emerge. So antispam filters must be updated continuously.

Data drift is often gradual but can reach an inflection point where model performance drops suddenly. The best defense is continuously retraining models on fresh, representative data. This allows the AI to keep pace with a changing world. Some techniques like synthetic data generation can also make models more robust. But ultimately, real-world data is required to keep systems adaptive.

CONTINUOUS MODEL TRAINING AND MONITORING

AI is not a one-time implementation but an ongoing journey. To avoid losing relevance and making poor decisions, continuous model training and monitoring are crucial. Companies must allocate resources specifically for this purpose, assigning personnel to monitor AI performance, flag areas of underperformance, and facilitate regular retraining with new data.

Employees working with AI systems should also be trained to identify situations warranting additional training data. AI maintenance should be built into budgets from the outset as neglecting this aspect can lead to gradual performance backsliding, potentially impacting business decisions and outcomes.

Consider a scenario where a model starts to misfire due to changes in the business environment since its initial training without any immediate recognition of the issue. The decisions made based on this AI model could adversely impact the business before the model aberration is caught. To mitigate such risks, models should be designed to evaluate themselves periodically, detect data drift, and inform the governance team when retraining is necessary to avoid unintended consequences of poor recommendations.

By addressing overfitting, underfitting, and data drift while implementing continuous model training and monitoring, organizations can develop robust and adaptive AI models that maintain their relevance and accuracy over time. This enables informed decision-making and drives sustained business value.

THE CHALLENGE OF ACCESSING DATA

In the previous sections we have discussed the issues associated with data quality and providing the right type of data to AI models. Another huge challenge for AI teams is accessing the data itself in many companies and organizations.

THE MAZE OF DATA SILOS IN ORGANIZATIONS

If a pharmaceutical company stored its research data in one department and patient data in another, these "data silos" could severely limit the potential of AI. If the company is developing a new drug, AI could analyze both research and patient data to predict the drug's effectiveness and potential side effects. However, if these datasets are siloed, the AI system can't provide a comprehensive analysis, leading to missed opportunities and inefficiencies.

As we discussed in the halicin discovery, the AI models had a 360 view of vast amounts of data in three categories—molecular structures of existing antibiotics, bacterial resistance to existing drugs, and chemical properties of various compounds. This access to data was a fundamental enabler of the discovery of halicin.[10]

Internal politics can often be a significant roadblock to effective data management and AI implementation. In many organizations, different departments or business units may be resistant to sharing data, often due to concerns about losing control or authority. This resistance can manifest in various ways—from reluctance to integrate data silos to outright refusal to participate in company-wide data initiatives. For example, a marketing team might be hesitant to share customer engagement data with the product development team, fearing it could influence budget allocations or strategic decisions. This lack of collaboration hampers the organization's ability to leverage AI fully, as the algorithms are only as good as the data they can access. Overcoming these internal barriers requires strong leadership and a culture shift toward data-driven decision-making, where the collective benefit of data sharing is recognized and rewarded.

Salesforce has tackled this issue head-on by integrating data from various sources into a single platform. Salesforce's Customer 360 platform pulls in data from sales, customer service, and marketing to offer a 360-degree view of the customer. This unified data model enables AI algorithms to analyze customer behavior more accurately, thereby enhancing decision-making.[11]

THE ROLE OF DATA GOVERNANCE

Data governance is another critical aspect of data management, especially when AI is involved. Misuse of data can lead to a loss of trust and hefty legal penalties, as seen in the backlash Facebook faced with the Cambridge Analytica scandal. In 2018, it was revealed that Cambridge Analytica, a political consulting firm, had harvested the personal data of millions of Facebook users without their consent. This data was then used for political advertising, violating user privacy on a massive scale. This misuse not only eroded public trust but also led to legal repercussions for Facebook, including a five-billion-dollar fine by the Federal Trade Commission.[12]

Companies must adhere to data privacy regulations like GDPR and ensure that data is collected and used ethically.

Transparency with users about how their data will be used can go a long way in building trust.

IBM has a robust data governance framework that ensures their AI systems not only have access to vast amounts of data but also use it responsibly, ethically, and securely.[13] Data governance policies should include strict access controls, regular audits, and compliance with data protection regulations like GDPR. This approach ensures that AI systems analyze large datasets to provide valuable insights without compromising data integrity or user privacy.

KEY ELEMENTS OF IBM'S DATA GOVERNANCE FRAMEWORK

- **Data Quality**—Make sure the data used is accurate and up to date. This is crucial for any AI system to function effectively.
- **Data Security**—Put in robust security measures to protect data from unauthorized access and cyber threats. This includes encryption, secure data storage, and stringent access controls.
- **Ethical Use of Data**—Have guidelines in place to ensure that data is used ethically. This includes not using data in a way that could be discriminatory or harmful.
- **Compliance**—Data governance framework should be compliant with various data protection regulations, such as GDPR in Europe or California Consumer Protection Act (CCPA). Some companies have dedicated teams that ensure ongoing compliance with these laws.
- **Data Accessibility**—While AI systems have access to vast amounts of data, the framework should ensure this data is used responsibly. It sets guidelines on who can access the data and for what purposes, ensuring no misuse.

- **Transparency**—Be transparent about how they use data. This includes clear communication with stakeholders and regular audits of their data usage practices.
- **Data Stewardship**—Appoint data stewards who are responsible for the quality and ethical use of data. These stewards oversee the data lifecycle, from collection to disposal, ensuring it aligns with IBM's governance guidelines.

By adhering to a robust data governance framework, you can ensure AI systems are not only effective but also ethical and secure. This commitment to data governance sets a standard many other companies aim to follow.

STRATEGIES TO ADDRESS DATA CHALLENGES

Effective use of data is paramount for organizations seeking to leverage the power of AI. Let's discuss strategies for addressing critical data challenges, emphasizing the importance of ensuring data quality as the foundation for trustworthy AI systems.

ENSURING DATA QUALITY: THE FOUNDATION OF TRUSTWORTHY AI

Data quality is the cornerstone of any AI system. Without it, even the most advanced algorithms can yield unreliable results. Uber relies on real-time data for fare calculation, route optimization, and passenger safety. A single incorrect GPS coordinate can lead to inefficient routes or safety risks. To mitigate this, Uber uses machine learning algorithms to automatically identify and correct data anomalies.[14] Similarly, EazyML detects data bias before machine learning models are trained, CEO Deepak Dube told me. These proactive measures ensure the trustworthiness of AI systems in making critical business decisions.

Having data stewards is critical to maintain the integrity of data. Alongside human expertise, leveraging machine learning algorithms for automated data cleaning can significantly enhance data quality. These algorithms can swiftly identify and rectify anomalies, reducing the burden on manual processes.

Lastly, regular audits are crucial. These assessments not only validate the data's accuracy but also its reliability, ensuring that the organization can trust the data for strategic decision-making. By combining human oversight with technological solutions and ongoing evaluations, business leaders can create a robust framework for maintaining high-quality data.

BROADENING THE DATA SPECTRUM: THE IMPORTANCE OF DIVERSITY

In 2015, Google's Photos app sparked controversy when it erroneously labeled a black couple as gorillas. This mistake led to widespread criticism of Google on social media due to the racist connotations of the label. Google promptly acknowledged the error with an executive calling it "100 percent not okay" and a high-priority bug. The company took immediate steps to prevent similar incidents and committed to long-term improvements in image recognition, particularly regarding dark-skinned faces and sensitive terminology.[15]

Diversity in data is not just an ethical obligation; it's a business necessity. This diversity not only enriches the data pool but also makes the resulting algorithms more robust and inclusive.

Another crucial aspect is bias assessment. Utilizing specialized tools to identify and mitigate biases in both data and algorithms can significantly improve decision-making and fairness. Finally, transparency is key. Being open about data practices not only builds trust among users but also positions the company as an ethical leader in the AI space.

SAFEGUARDING USER PRIVACY: BALANCING UTILITY AND ETHICS

Data privacy is a critical concern in today's digital landscape. Apple sets a high bar by using differential privacy, allowing the company to collect user data without compromising individual privacy.[16] For leaders looking to emulate such best practices, several strategies can be employed.

First, consider implementing differential privacy techniques similar to those used by Apple. Differential privacy is a mathematical method that ensures privacy by making minor alterations to individual data in a group, thereby revealing group patterns without disclosing specific individual information. This allows for valuable data analysis while ensuring user privacy remains intact.

Second, always prioritize obtaining user consent before collecting and using their data. This not only adheres to legal requirements but also fosters trust between the company and its users. Lastly, employ strong encryption methods to safeguard user data from unauthorized access and potential breaches.

RIGOROUS AI TESTING: ENSURING RELIABILITY AND SAFETY

By adopting a multipronged strategy involving comprehensive scenario testing, continuous refinements based on real-world data, user feedback, and robust safety protocols, business leaders can significantly enhance the reliability and safety of their AI systems. Rigorous testing is vital to build AI that is not only powerful but also trustworthy.

Tesla's autopilot system is a prime example of an AI system that undergoes extensive real-world testing. They have set up a seventy-mile driving loop around their Michigan headquarters to track the self-driving technology over time. The autopilot system is tested in various real-world scenarios, including navigating obstacle-laden parking lots, executing U-turns, and intricate left turns. Based on the results of these tests, Tesla continuously refines its algorithms.[17]

When testing AI systems, it's crucial to simulate real-world scenarios. Black box testing techniques for AI involve simulating AI logic outcomes over a specific timeframe and comparing them with real-world results.[18] This helps assess how the AI performs in diverse conditions.

Tesla's approach to refining its autopilot system exemplifies continuous improvement based on lessons learned from real-world use. They make ongoing algorithm enhancements leveraging data from extensive testing.

Ensuring safety is paramount when developing AI systems. Anthropic adopted Responsible Scaling Policy (RSP), which outlines technical and organizational protocols aimed at managing the risks associated with developing advanced AI systems. The policy categorizes AI systems by their potential for catastrophic risk, requiring varying degrees of safety and security measures based on these levels.[19] Similarly, the US Artificial Intelligence Safety Institute (USAISI) at The National Institute of Standards and Technology (NIST) brings together over two hundred organizations to develop science-based guidelines and standards for reliable and safe AI measurement and deployment.[20]

LEVERAGING CLOUD FOR ENHANCED DATA MANAGEMENT

Cloud-based data management allows organizations to store and process data in a scalable, flexible, and secure way without the need for expensive hardware investments. It provides robust data backup and recovery mechanisms, ensuring data protection against loss or corruption. Cloud services offer advanced analytics and data processing tools, enabling efficient insights from data. Additionally, cloud-based data management facilitates remote access and collaboration, promoting better data sharing and data-driven decision-making.

CHAPTER 7

An Ounce of Prevention: Securing the Competitive Edge of AI

On May 22, 2023, social media was ablaze with a photograph showing an explosion near the Pentagon. First posted on Facebook and then amplified by influential Twitter accounts like RT and ZeroHedge, the photo sent shockwaves through the public and even caused the S&P 500 index to dip nearly fifty points within hours. However, the photo was a hoax. This image wasn't just any photo but the creation of an AI model capable of creating hyperrealistic images from text prompts. The Pentagon and the Arlington Police Department, along with open-source intelligence researchers, quickly debunked the image by pointing out inconsistencies such as the building's shape and the crowd barriers.[1] Despite the quick rebound of the stock market and corrections posted by those who had shared the image, the damage—both psychological and financial—had been done.

This incident serves as a cautionary tale. It underscores the urgent need for security measures and secure AI technologies ranging from digital watermarking to steganography. As AI models become increasingly sophisticated, the line between reality and AI-generated content blurs, posing significant risks to public trust, market stability, and even national security.

The allure of AI lies in its ability to process vast amounts of data, learn from it, and make decisions. But this strength is also its vulnerability. Unlike traditional software that follows explicit instructions, AI models learn from data. This learning process can be exploited or misled.

Think of AI security as the lock for your treasure chest of AI assets—data, algorithms, and infrastructure. It's not just about keeping intruders out; it's also about ensuring the treasure inside remains intact and authentic.

WHY AI SECURITY MATTERS

The growing reliance on AI—from supply chain optimization to predictive analytics—exposes businesses to unique security risks that leaders cannot afford to overlook. Lax AI security puts business continuity, intellectual property, and customer trust at risk. That's why every business leader should prioritize identifying and addressing potential weak spots—from flawed data to vulnerabilities in third-party AI software.

MINIMIZING BUSINESS DISRUPTION

AI systems are increasingly used for mission-critical tasks like fraud detection and demand forecasting. A breach that compromises these systems can wreak havoc on business workflows. In July 2021, a devastating ransomware attack crippled over fifteen hundred businesses that relied on AI-driven network management software from Kaseya. The REvil cybercriminal gang encrypted the victims' data and demanded seventy million dollars in cryptocurrency to decrypt it. Many affected businesses suffered severe disruptions, unable to access critical systems and data.[2] Proactively securing AI systems reduces the likelihood of such disruptive breaches.

SAFEGUARDING INTELLECTUAL PROPERTY

For many companies, their proprietary AI algorithms and models are a core competitive advantage worth protecting. An attacker who gained access to these AI assets could duplicate or even improve upon the technology. IBM reported that 32 percent of all cyberattacks were attempts to steal data and 50 percent of the attacks were focused on attacking AI technologies.[3] AI systems, which are complex and distributed, are inherently prone to security vulnerabilities due to their intricacy. Additionally, AI's reliance on large, often sensitive datasets for training introduces significant security challenges, especially when this data is in transit or stored in cloud environments. Moreover, the rapid pace of AI development can compel companies to compromise on security measures as they quickly push out new products and features. Robust AI security is key to safeguarding this intellectual property from theft or misuse.

MAINTAINING CUSTOMER TRUST

Customers expect their sensitive data to be handled securely. A breach exposing customer data can severely erode public trust in a company. We saw this play out when a massive data breach in 2017 compromised personal information of over 145 million Equifax customers. The exposed data included names, social security numbers, birth dates, and addresses. This represented nearly half the US population. The backlash was immense, and the company spent $1.4 billion in cleanup costs and Moody's downgraded the company's financial rating.[4]

TYPES OF THREATS TARGETING AI SYSTEMS

The expected adoption of AI systems across companies exposes them to a range of sophisticated threats. Understanding these threats is essential for safeguarding AI systems and ensuring their reliability and security. Let's explore the primary types of threats targeting AI systems, highlighting their mechanisms and potential impacts. By examining real-world examples and the methodologies behind these attacks, we can better prepare to defend against them and protect the integrity of AI-driven innovations.

ADVERSARIAL ATTACKS

In 2012, AlexNet, a convolutional neural network (CNN), a specialized type of deep learning algorithm designed for object recognition, won ImageNet's challenge of recognizing images. It did so with an amazing 85 percent accuracy. But deep neural networks (DNN) like AlexNet are easily fooled. Researchers took a correctly classified photo by AlexNet and made minor pixel changes. To human beings, the changed image looked identical to the previous photo, but AlexNet classified a temple as an ostrich with high confidence.[5]

This is an example of an adversarial attack, which involves manipulating the input data to an AI model in such a way that the model makes a mistake, even though the change is almost imperceptible to humans. These attacks occur during the inference stage, which is when the AI model is already trained and deployed. The aim is to trick the AI into making a specific mistake by altering the input data. The model itself remains unchanged. A self-driving car's AI system is designed to recognize stop signs. An attacker could subtly alter the stop sign's appearance in a way that's almost imperceptible to humans but enough to make the AI misinterpret it, potentially causing an accident. Adversarial attacks can be applied to deep neural networks that are focused on text, speech, or other forms as well. The consequences of such actions are significant. In text-based systems, adversarial inputs could lead to the generation of harmful or misleading information. In speech recognition systems, they might cause misinterpretations that lead to incorrect responses or actions, impacting areas such as voice-activated banking or security systems. The implications are broad, ranging from minor inconveniences to severe safety risks and financial losses, highlighting the critical need for robust defenses against such vulnerabilities in AI systems.

DATA POISONING

Data poisoning means tampering with the training data of an AI model, thereby affecting its learning process and subsequent performance. These attacks happen during the training stage, affecting how the AI model learns. The goal is to corrupt the training data so the AI model itself becomes flawed, affecting all subsequent inferences. Consider a recommendation engine for an online retail store. If someone manipulates the training data to include false preferences, the AI could start recommending irrelevant or inappropriate products, affecting customer satisfaction and sales.

In 2016, Microsoft released a Twitter chatbot called Tay. Malicious users quickly began feeding offensive tweets to Tay, which had been trained on a massive dataset of text and code. This caused Tay to start posting racist and sexist tweets, and Microsoft was forced to shut it down within hours of its launch.[6] Researchers at Tencent showed how they were able to trick the windshield wipers on a Tesla's self-driving car's AI to activate by using a hidden pattern on a TV screen.[7]

MODEL TAMPERING

Model tampering is another critical security concern that directly targets the AI model's architecture and parameters. Unlike data poisoning, which corrupts the model indirectly through its training data, model tampering involves unauthorized changes to the model itself after it has been trained. Let's consider a credit scoring AI model that banks use to determine loan eligibility. If someone with malicious intent gains access and alters the model's parameters, they could manipulate the system to approve loans for unqualified applicants or deny loans to deserving candidates. The objective is often to degrade the model's performance or make it behave in a specific, malicious way.

Researchers showcased how an attacker could manipulate the parameters of a machine learning model designed for malware detection, leading it to incorrectly classify malware as safe. By subtly changing the model's weights, they reduced its accuracy from over 99 percent to about 3.5 percent. A study demonstrated that tampering with the parameters of language models like GPT-2 after training is feasible, allowing for the injection of specific biases or the generation of offensive outputs for selected prompts. The researchers achieved this targeted misbehavior by altering only three neurons in the model.[8]

Such tampering can have far-reaching consequences, affecting a company's bottom line and customer trust. On a separate example, researchers in Israel demonstrated how facial recognition models could be tampered with to impersonate certain individuals. By modifying the model's parameters, they generated false positives for a targeted person even on images that did not contain their face.[9] This raises alarming concerns about tampering risks with biometric AI systems.

INFRASTRUCTURE ATTACKS

Infrastructure attacks target the hardware and software platforms where AI models are hosted, aiming to compromise the entire AI system. Picture a cloud-based customer service chatbot. If the cloud infrastructure is compromised, not only could sensitive customer data be exposed, but the chatbot could also start providing incorrect or misleading information. In a different scenario, if an attacker gains control over the server hosting a health care AI system, they could alter diagnostic algorithms, leading to incorrect medical advice or prescriptions.

In 2021, malicious actors breached Nvidia's infrastructure, stealing sensitive data related to AI projects. Exploiting network vulnerabilities, the attackers infiltrated the system and exposed critical insights into Nvidia's advanced AI research and chip architectures. This breach highlighted the complexity of modern cyber threats, as they bypassed security measures to access proprietary information. The theft provided the attackers with an in-depth understanding of the company's state-of-the-art technologies, potentially jeopardizing the company's competitive edge and disrupting future innovations.[10]

In 2015, a German steel mill suffered extensive damage when hackers infiltrated its industrial control systems and prevented a blast furnace from shutting down properly. The attackers used a sophisticated phishing campaign to trick employees into downloading malicious software, which gave them control over the plant's operational technology network. By exploiting the AI systems used to optimize steel production, the hackers furthered their control over the process. This led to significant physical damage, costly repairs, and operational downtime due to the overheating furnace. The incident highlighted the vulnerabilities in industrial control systems and the critical need for robust cybersecurity measures to protect AI-driven industrial processes.[11]

Both incidents exemplify how infrastructure attacks that disrupt the backend systems supporting AI can enable theft of confidential AI data while also leading to physical damage of equipment. Hardening their infrastructure security and having robust incident response plans are crucial to counter such cyber threats targeting their AI assets and operations.

EAVESDROPPING AND DATA INTERCEPTION

Eavesdropping and data interception involves unauthorized access to data during its transmission, either during the training phase or when the AI model is making inferences. The focus is on unauthorized access to data as it's being transmitted but not on altering the AI model or its inputs. The data is "listened to" or captured for malicious use. Let's say you're using AI to analyze proprietary chemical formulas. If someone intercepts this data during the analysis, they could potentially steal or replicate your formulas, leading to significant business losses.

AI systems are vulnerable to eavesdropping attacks that intercept sensitive data used to train or operate them as unencrypted data creates vulnerability. In 2015, researchers from Carnegie Mellon University demonstrated how voice assistants like Siri could be attacked using ultrasonic tones to secretly record conversations with users.[12] This highlighted privacy risks from potential audio data interception. In 2020, threat actors compromised Tesla's data transmission systems to gain access to driving data collected from Tesla vehicles. The hackers got insights into Tesla's AI training data used for self-driving technology.[13] These examples underscore the importance of encrypting data flows and having robust cybersecurity to safeguard proprietary training data and operational inputs for AI systems against interception attempts.

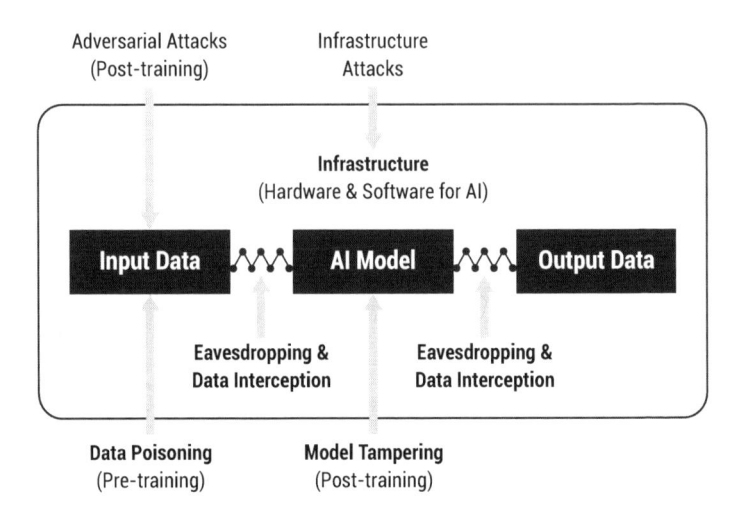

Fig 9: Types of Threats Targeting AI Systems

CORE COMPONENTS OF AI SECURITY

Robust AI security requires a wide lens on the big picture and a magnifying glass on the finer details. Let's break down the core components of AI security.

DATA INTEGRITY: THE STURDY FOUNDATION

Data integrity is the bedrock for reliable AI systems. This means verifying the accuracy and consistency of the data used to build and run AI models. Data corruption destabilizes the entire AI structure. For example, a package delivery AI relies on mapping data. If this data was hacked and street names altered, the AI could create extremely inefficient routes. Safeguarding data integrity should involve encryption and access controls to prevent unauthorized changes, ongoing data validation through checksums and error detection, and diligent data hygiene procedures like input sanitization. With pristine data as a solid baseline, companies can confidently unleash AI to drive decisions and automate processes without skewed results.

TRANSPARENT ALGORITHMS: OPENING THE BLACK BOX

In addition to data integrity, companies need transparency into how their AI systems operate—shedding light inside the AI black box. Algorithmic transparency means understanding how the AI reaches conclusions or takes actions based on its training. An AI screening job candidates could discriminate unfairly, and transparency of its decision-making logic is critical to correct it. Transparency enables both ethical oversight and calibration of the AI to align with business goals.

Companies can audit algorithms by documenting the data sources, assumptions, and computational steps, building model explainability through techniques like LIME (local interpretable model-agnostic explanations) and SHAP (SHapley Additive exPlanations) and conducting bias testing and correction throughout development. LIME is used to explain the predictions of any machine learning classifier. It works by perturbing the input data and observing the resulting changes in the prediction to identify which parts of the input are most influential. This technique allows for a better understanding of why a model made a particular decision.

SHAP is a method based on game theory that provides consistent and accurate feature importance values for any machine learning model. SHAP values quantify the contribution of each feature to the prediction by distributing the prediction difference among the features. This helps in identifying the impact of each feature on the model's output, making the model's behavior more interpretable and understandable. By transparently conveying the AI's rationale, companies build stakeholder trust and prevent harmful "black box" effects.

ACCESS CONTROL: PERMITTING AUTHORIZED ACCESS

Access controls limit which personnel can interact with the AI system and its data, like having ID card access to rooms in a corporate building. Proper controls prevent both insider misuse and external attacks. For example, marketers optimizing an AI recommendation engine should not access user data without permissions and auditing. A well-designed access protocol specifies roles, permissions, and access tiers, uses secure protocols like SSH keys for data access, implements multifactor authentication for control panels, and logs activity for regular audits. These layered controls lock down the AI against breaches while enabling controlled access for authorized users.

CONTINUOUS MONITORING: THE TWENTY-FOUR-SEVEN SECURITY CAMERA

AI systems require constant vigilance, like a bank using security cameras to monitor crimes. Continuously tracking performance, behavior, and access patterns allows prompt detection of any anomalies or attacks. Ongoing monitoring should include: tools to analyze system logs, network traffic, and access logs for threats, model performance metrics to detect any degradation, dashboards for easy visualization of all critical security parameters, and alerts for authorized teams about suspicious activities. With holistic monitoring, issues can be identified early before they spiral out of control.

REGULATORY COMPLIANCE: FOLLOWING THE RULES OF THE ROAD

Much like drivers following traffic laws, companies deploying AI must adhere to relevant regulations and ethical codes. These include guidelines on data privacy, algorithmic fairness, responsible AI practices, and industry-specific laws. Compliance involves: regular audits by internal/external oversight bodies, documentation of policies and controls demonstrating due diligence, training programs for developers on ethical AI practices, and impact assessments before AI deployment. Compliance protects organizations from legal penalties and reputational damages. But more importantly, it ensures AI is developed and used responsibly.

BEST PRACTICES FOR SECURING AI SYSTEMS

Here are some best practices for securing AI systems.

REGULAR SECURITY AUDITS

A security audit is a systematic evaluation of the security of your AI system. It identifies vulnerabilities and assesses how well the system can defend against potential threats. Conducting frequent security audits can help identify vulnerabilities in your AI systems before they can be exploited.

Here are the key steps to conduct a security audit:

- **Scope Definition:** Determine what parts of the AI system will be audited. This could be any stage, the data, the model, or the entire pipeline.
- **Risk Assessment:** Identify potential vulnerabilities that could affect the AI system.

- **Tool Selection:** Choose appropriate auditing tools such as IBM's Adversarial Robustness Toolbox for AI.
- **Audit Execution:** Run the audit using the selected tools and methodologies.
- **Report and Remediation:** Document the findings and develop a plan to address identified vulnerabilities.

DATA ENCRYPTION

Data encryption converts data into a code to prevent unauthorized access. Encrypting sensitive data is crucial, especially for AI systems that handle confidential information. Determine what data needs to be encrypted. Opt for strong encryption algorithms like AES-256. Use encryption libraries or services to encrypt the data before it's used in the AI system. Advanced techniques like homomorphic encryption allow computations on encrypted data, providing an extra layer of security.

MULTIFACTOR AUTHENTICATION (MFA)

Implementing MFA adds an additional layer of security by requiring two or more verification methods—a password, a smart card, or biometric verification like a fingerprint. Choose an MFA solution that integrates with your existing systems. Set up policies to determine when MFA is required. Educate employees on how to use MFA and why it's important.

ROLE-BASED ACCESS CONTROL

Limiting access to AI models and data through role-based access control can mitigate the risk of internal threats. Only authorized personnel should have the ability to view or modify critical AI components. Identify different roles within the organization and what level of access they should have.

Configure the AI system to restrict access based on these roles. Periodically review and update roles and permissions to adapt to organizational changes.

SOFTWARE UPDATES

Keeping software up to date is essential. Regular updates and patches can fix vulnerabilities that could otherwise be exploited by attackers.

EMPLOYEE TRAINING

Employees often serve as the first line of defense against security threats. Training programs that educate staff on the importance of AI security and how to recognize potential threats can be invaluable. Create a training program that covers essential AI security topics. Use workshops, webinars, or e-learning platforms to deliver the training. Keep the training updated to include the latest security threats and best practices.

TOOLS AND TECHNOLOGIES FOR AI SECURITY

Here's a rundown of some essential tools and technologies that can enhance AI security—both traditional and emerging.

TRADITIONAL TOOLS

Firewall Solutions—Firewalls protect your AI system from potential threats from the internet. Choose a solution that offers deep packet inspection and configure it to block unauthorized access to your AI servers.

Intrusion Detection System (IDS)—IDS monitors network traffic for suspicious activities and issues alerts when potential threats are detected. Implement IDS solutions that offer real-time monitoring and integrate them with your existing security infrastructure.

Data Encryption Tools—These tools encrypt sensitive data, making it unreadable to unauthorized users. Opt for encryption tools that offer strong algorithms like AES-256, and implement them to secure data at rest and in transit.

Secure Containers—Containers provide an isolated environment for running your AI applications and enhancing security. Use containerization platforms like Docker to package your AI applications along with all its dependencies, ensuring a secure runtime environment.

Adversarial Robustness Toolbox—This is a specialized tool for evaluating the security of AI models against adversarial attacks. Integrate the toolbox into AI development pipeline to assess model robustness and identify vulnerabilities.

Zero Trust Architecture—Zero trust architecture operates on the principle of "never trust, always verify," ensuring stringent access controls. Implement zero trust principles by enforcing strict access policies and continuous authentication for all users and devices.

EMERGING TECHNOLOGIES
STEGANOGRAPHY WATERMARKING

Steganography watermarking allows AI models to be marked to deter unauthorized copying or usage. It acts like a digital signature, ensuring traceability and protecting intellectual property rights. Implement watermarking techniques during the model training phase to embed a unique identifier into the AI model.

I had the opportunity to work with Wendy Chin, founder and CEO of PureCipher, an AI security start-up. The company is a pioneer in watermarking with its Universal Multiplex Watermark (UMW) for data integrity solution, which marks all types of digital data like images, plaintext, audio, video, and even executables. For text files like PDF, it uses a patent-pending technology to embed encrypted user-defined information directly in the file that are not human readable. This enables provenance and lineage tracking while allowing tamper detection to avoid data poisoning. For visual and audio files, PureCipher uses steganography technique to conceal encrypted watermarks inside the existing images or audio. These hidden watermarks can contain verification data or even entire separate files embedded invisibly.

The key benefit of this technology is fourfold. It allows proving authenticity of files like X-rays and patient records without changing their visible contents. It enables tracking the origin and modifications made to medical data to detect tampering. Watermarking is invisible, so workflows are unaffected for doctors and patients. And finally, it provides machine-readable security without the need for extra hardware, only software.

Google has launched an invisible, permanent watermark on images to identify them as computer-generated, aiming to prevent the spread of misinformation.[14] Adobe and IBM have signed up to a voluntary AI governance scheme that includes watermarking AI-generated content.[15]

COMPUTE ON ENCRYPTED DATA

Homomorphic encryption is a groundbreaking technology that allows AI models to work directly on encrypted data without ever decrypting it. Integrate homomorphic encryption libraries into your AI system to ensure data remains confidential even during computation.

Microsoft's Simple Encrypted Arithmetic Library (SEAL) aims to make homomorphic encryption easy to use and available for everyone.[16] PureCipher uses fully homomorphic encryption to enable foolproof security at scale. Duality, another start-up uses homomorphic encryption to build more privacy-centric data collaboration tools.[17]

OTHER NOTEWORTHY TECHNOLOGIES

In addition to watermarking techniques like UMW, cutting-edge cryptographic methods are also being explored to further secure AI systems. Secure multi-party computation allows multiple entities to jointly analyze data or train models without exposing their private data.[18] Differential privacy adds noise to model training to prevent leaking individual data points.[19] Zero-knowledge proofs can cryptographically verify statements without revealing sensitive details.[20]

For example, an AI model could prove it was trained on high-quality datasets without exposing the actual data. Such advanced privacy-preserving approaches enable collaborations and transactions involving AI or data exchange with enhanced trust. As AI becomes further ingrained in sensitive domains like finance, security, and health care, techniques that strengthen data confidentiality and integrity without compromising utility will be increasingly critical.

The promise of AI comes hand in hand with new threats. As AI enters business operations, securing these mission-critical systems is imperative. With vigilance, foresight, and a proactive cybersecurity strategy, the risks can be managed. With a holistic view encompassing data, algorithms, and infrastructure, security can transition from an afterthought to a core competitive advantage.

CHAPTER 8

The Art of AI Leadership: Building AI Teams

"If you are a CEO, you can't just say, I am going to get my tech guys to understand it and educate me. You must understand it because it will have significant impact on every single thing you do." — Mark Cuban on a CNBC interview, March 4, 2024.

The emergence of artificial intelligence is transforming businesses in unprecedented ways. As AI capabilities grow more sophisticated, companies must adapt their leadership strategies and teams to harness their full potential. This chapter offers invaluable guidance to business leaders on cultivating the mindsets, skills, and diverse teams needed to spearhead AI adoption. It emphasizes the human touch required to ethically align AI with business goals and societal values. For companies seeking to gain a competitive edge through AI, the insights here chart a path forward.

LEADERSHIP IN THE AGE OF AI: A MULTIDIMENSIONAL APPROACH

The rise of artificial intelligence (AI) has ushered in a profound shift in the leadership landscape. In this era of rapid technological advancement, organizations can no longer rely solely on traditional managerial approaches. Instead, they require visionary leaders who can navigate the uncharted territory of AI adoption and harness its transformative potential.

VISIONARY AND STRATEGIC GUIDES

Leadership in the AI age demands a fundamental transition from task-oriented management to visionary guidance. Leaders must bridge the gap between technical experts and business stakeholders, acting as interpreters who translate complex AI concepts into actionable strategies. Possessing a deep understanding of AI technologies, and other technologies such as cloud computing is essential, but it must be coupled with a comprehensive grasp of the organization's core business model, challenges, and opportunities.

While addressing immediate challenges is essential, effective AI leaders must also maintain a future-focused vision. This involves anticipating AI trends and charting a clear roadmap for the organization's long-term AI journey. By aligning short-term tactics with strategic long-term goals, leaders ensure their organizations leverage AI's full potential for sustainable success.

AI leaders are faced with critical decisions, such as whether to build AI solutions in-house or outsource to specialized vendors. They must skillfully evaluate the merits and trade-offs of each approach, factoring in considerations like cost, implementation speed, and expertise. Moreover, they must adeptly manage relationships with vendors, ensuring they meet contractual obligations, compliance standards, and quality benchmarks.

CHAMPIONS OF TRANSFORMATIONAL CHANGE

The integration of AI disrupts traditional structures, workflows, and the nature of work itself. Leaders must champion change management, guiding teams through these disruptions while fostering a culture that embraces AI's potential. Transparency is key, demystifying processes, setting clear expectations, and fostering accountability.

While technology and data are critical AI project components, the human factor is often overlooked. Resistance to change among employees can significantly impede successful AI integration. According to McKinsey & Company, 70 percent of transformations fail, with a key reason being resistance to change and inadequate management support.[1]

Akin to paddling upstream, even the strongest technological current will struggle if the rowers (employees) are not synchronized or actively resisting. Resistance is a natural human response, particularly to disruptive technologies like AI, stemming from fears of job losses, irrelevance, or the unknown. This resistance can manifest as lack of engagement, poor adoption rates, or active sabotage of new systems.

To overcome resistance, engaging employees throughout the change process is crucial. This involves transparent communication about the AI project's scope, its impact on their roles, and measures to mitigate negative effects like job displacement. Including employees in decision-making can empower them and increase their likelihood of supporting the change. Effective training and transition programs aimed at upskilling employees and preparing them for AI-driven changes can combat resistance. For example, manual quality control staff could be trained to manage and interpret AI system outputs, shifting their roles rather than eliminating them.

ELEVATING THE HUMAN ELEMENT

Educating employees on how AI augments and complements human capabilities, rather than replacing them, is essential. Leaders with high emotional intelligence (EQ) can empathize with their teams' concerns and feelings, ensuring the human element remains central to AI-driven initiatives. They understand that while AI-driven efficiencies may lead to workforce reductions, such decisions must be weighed against the broader impact on team morale, well-being, and organizational culture.

A leader with high EQ would approach such a scenario with compassion and sensitivity. They would consider the emotional toll of workforce reductions on their team members, the potential loss of institutional knowledge and expertise, and the potential damage to the organizational culture fostered over years of collaboration and trust-building.

Moreover, the rapid pace of AI development demands a culture of continuous learning and adaptability. Leaders who invest in training and development programs empower their teams to embrace AI advancements proactively. By equipping employees with the necessary skills and knowledge, leaders foster a sense of resilience and confidence, enabling their teams to thrive in an AI-driven future.

Furthermore, actively engaging with the global AI community allows leaders to glean valuable insights, best practices, and anticipate emerging trends. However, this engagement should not be solely focused on technological advancements; it should also encompass discussions on the ethical considerations, societal implications, and the preservation of human values in an AI-driven world.

Ultimately, while AI holds immense potential for innovation and efficiency, the human element—our emotional intelligence, creativity, and ability to empathize—will remain the driving force behind successful organizations. Leaders who prioritize the human aspect, while strategically leveraging AI capabilities, will be best positioned to navigate the challenges and opportunities presented by this technological shift.

CATALYSTS FOR CROSS-FUNCTIONAL ENGAGEMENT

Steering AI-driven transformation requires a symphonic approach, where leaders harmonize the diverse talents and perspectives within their organizations. Building high-performing AI teams is an artform one that demands curating a rich blend of experiences, backgrounds, and worldviews. This multidimensional diversity transcends academic and professional boundaries; it encompasses the nuanced cultural lenses that are pivotal for crafting AI solutions with global resonance.

To unlock peak synergies, leaders must be maestros of collaboration, choreographing seamless workflows across functions and roles. This interplay extends beyond simply aligning technical and commercial elements. It necessitates integrating perpetual feedback loops that propel continuous evolution and refinement. Upskilling pathways and professional growth plans empower team members to the cadence of AI.

In this work, an empowering culture is the source of inspiration. Leaders must foster an environment where everyone can contribute their unique ideas. Initiatives like cross-functional ideation sessions, collaborative sprints, and innovation incubators can strike this powerful chord. By harnessing their organizational choir's harmonic potential, leaders can compose AI masterworks that are both pioneering and aligned with overarching business themes.

ETHICAL AND LEGAL GUARDIANS

The integration of AI presents a myriad of ethical dilemmas—from data privacy concerns to the potential for algorithmic biases. Leaders must serve as the organization's moral compass, prioritizing ethical considerations over short-term gains. For instance, an ethical leader would refrain from exploiting vulnerable populations through AI-driven marketing strategies, even if it promised financial rewards.

Leaders must also stay abreast of the evolving legal landscape surrounding AI, including data protection laws, intellectual property rights, and industry-specific regulations. Ensuring compliance with these guidelines is not only a legal obligation but also a means of building trust with stakeholders. This involves implementing robust data governance frameworks, conducting regular audits, and maintaining transparency about how data is collected, used, and stored.

Moreover, leaders should advocate for the responsible use of AI within their organizations. This includes promoting fairness, accountability, and transparency in AI systems and ensuring that ethical considerations are embedded in every stage of AI development and deployment. By doing so, leaders can mitigate risks, enhance trust, and foster a culture of ethical AI practice.

MEASURING PROGRESS, DRIVING SUCCESS

As with any transformative technology, establishing key performance indicators (KPIs) is crucial for measuring the success of AI initiatives. These metrics could range from cost savings and revenue growth to customer satisfaction scores. Regular review of these metrics ensures alignment between the team's efforts and business objectives, providing data-driven insights to inform future AI projects.

Leaders should adopt a balanced scorecard approach to measure progress, incorporating financial, operational, customer, and innovation metrics. This holistic view allows for a comprehensive assessment of AI's impact on the organization. Additionally, leaders should set clear, measurable objectives upfront to align the team toward a unified vision and adopt agile methodologies to enhance flexibility amid the dynamic nature of AI work.

Anticipating and resolving conflicts through open dialogue is another critical aspect of measuring progress. Leaders should leverage conflicts as opportunities for innovative solutions, fostering a culture of continuous improvement. By regularly evaluating the effectiveness of AI initiatives and making necessary adjustments, leaders can ensure that their organizations remain agile and responsive to changing market conditions.

BUILDING AN AI DREAM TEAM: KEY PLAYERS FOR SUCCESS

Success of AI across companies and organizations often hinges on the collective strengths of a diverse team. Each member brings a unique skill set, ensuring that the AI project is well-rounded, innovative, and effective. Let's explore the essential roles that should be part of any robust AI team.

Chief AI Officer: The chief AI officer is a visionary leader responsible for driving an organization's AI strategy and spearheading its adoption of AI technologies. This cross-functional leader must possess a unique blend of technological expertise, business acumen, leadership skills, and ethical awareness to navigate the complexities of the AI era and unlock its full potential.

Data Scientists: Often considered the backbone of any AI project, data scientists possess expertise in statistical analysis and use these skills to interpret and understand complex data structures. They develop algorithms, predictive models, and machine learning strategies to extract insights and information from data.

Data Engineers: While data scientists focus on analysis, data engineers handle the architecture and infrastructure. They design, construct, install, and maintain large-scale processing systems and other infrastructure. Their role ensures the data is clean, reliable, and easily accessible.

Machine Learning Engineers: These specialists take the prototypes of models developed by data scientists and turn them into working models that can be integrated into production systems. They have a deep knowledge of programming and machine learning algorithms.

Software Engineers: While data scientists build the models, software engineers productionize and integrate the models into products and applications. They bring valuable software skills like system design, coding, testing and maintenance.

UI/UX Designers: Design is crucial because AI systems have an end-user. UI/UX designers optimize the interaction between humans and the AI product. They make the output interpretable and actionable.

Domain Experts: Since AI models need to be tailored to specific domains like health care and finance, it's crucial to have team members with in-depth knowledge of the domain. They can provide insight into the key problems, datasets, and evaluation metrics.

Ethics Experts: As AI systems grow more powerful, ethics oversight becomes necessary to address bias, privacy, security, and social impact. Ethics experts act as the moral compass of the team, ensuring ethical considerations are not an afterthought but an integral part of AI development.

Product Managers: By acting as a bridge between technical AI capabilities and user/customer needs, product managers are essential in driving successful AI-led business transformations. They ensure that AI technologies are integrated seamlessly into products and services, delivering value to users/customers while aligning with the organization's overall AI strategy and ethical principles.

Project Managers: Managing the development, testing, and deployment of AI systems requires outstanding project management skills. This includes coordinating across teams, setting timelines and milestones, and communicating progress.

Now let's look into a real-world example. At Lightful, a capacity-building organization for nonprofits, the need to integrate generative AI technologies into their products and services prompted the formation of a cross-functional "AI Squad." This team was comprised of designers, digital coaches, engineers, and product experts, each bringing their unique expertise to the table. The diverse perspectives facilitated user-centric design, collaborative iteration, and a holistic understanding of AI's capabilities and limitations. The AI Squad's cross-functional nature proved invaluable in navigating the complexities of AI development. Through practices like user story mapping, continuous prompt testing, and cross-collaborative feedback loops, the team could rapidly iterate and refine AI prototypes. Moreover, the squad's varied backgrounds allowed for ethical considerations, potential biases, and societal impacts to be addressed, ensuring responsible AI deployment. Ultimately, the AI Squad's collaborative approach enabled Lightful to successfully integrate AI capabilities, such as an "AI Feedback" feature for social media post optimization, into their existing products, delivering value to their users.[2]

CRITICAL SKILLS FOR AI

AI projects have unique demands compared to traditional software projects. While both require an overlap in skills, certain skills are particularly critical for AI endeavors. Let's break down the skills that are especially vital for AI projects compared to other software projects.

While leaders play a crucial role in translating technical jargon into business strategy, it's equally important for AI teams to have a basic understanding of business objectives. This ensures the solutions they develop are aligned with the company's goals. For example, a data scientist should understand how their predictive model impacts not just user experience but also the bottom line.

Algorithmic Knowledge: A deep grounding in machine learning algorithms, neural networks, and mathematical disciplines like linear algebra and statistics is essential for AI due to the complexity and variety of algorithms required.

Data Management: Skilled data handling, preprocessing, cleaning, and comprehension of large datasets is critical for AI, especially machine learning approaches that thrive on data availability.

Programming Proficiency with AI Frameworks: Unlike traditional software projects, AI necessitates familiarity with specialized AI frameworks like TensorFlow, Keras and PyTorch.

Domain Expertise: AI projects benefit greatly from a nuanced understanding of the specific domain, such as medical terminologies for health-care AI, to enhance model effectiveness.

Ethical Considerations: AI raises unique ethical challenges around bias, transparency, and fairness that have profound implications due to the decision-making capabilities of AI systems.

Research Acumen: Staying abreast of the latest AI methodologies, algorithms, and best practices is paramount in this rapidly evolving field, much more so than in stable, traditional software domains.

UPSKILLING EMPLOYEES IN THE AI AGE TO BRIDGE THE TALENT GAP

AI is rapidly transforming the workforce, with the World Economic Forum estimating that AI could displace a significant portion of activities across many occupations, resulting in the loss of over 85 million jobs by 2025.[3] Yet even amid this looming unemployment crisis, companies already report shortages for skilled talent to develop and manage new AI systems. This growing mismatch between human skills and those needed to thrive in an AI-powered world signals an urgent need for businesses to invest in reskilling and upskilling their workforces.

INADEQUATE TALENT POOL

Having the right talent is as crucial as having quality data or clear objectives. Michael Jordan once said, "Talent wins games, but teamwork and intelligence wins championships."[4]

Without a skilled team, even the most well-funded AI initiative can falter. Successful AI projects require cross-functional teams that include business strategists, domain experts, and ethical advisors to ensure the solution is technically sound, aligned with business goals, and ethically responsible.

Consider a hypothetical scenario where a FinTech company hires data scientists fresh out of academia for a fraud detection project. Although their algorithms are mathematically sound, they lack industry-specific insights. This gap in knowledge results in a high rate of false positives, ultimately leading to the project's failure.

The skills gap extends beyond technical expertise to understanding how to integrate AI into business processes and strategies. Bridging this gap requires investing in training programs to upskill the existing workforce and fostering a culture of interdepartmental collaboration.

THE SKILLS GAP WIDENS

On one hand, AI and automation are making many jobs such as customer service specialists redundant. Simultaneously, these technologies are creating new specialized roles like AI researchers, data analysts, machine learning engineers and automation specialists. Positions requiring technical expertise to build, implement and manage AI systems are seeing massive demand. But talent supply is scarce even amid swelling unemployment.

This bifurcation is creating a significant skills gap across industries. Employers must decide between investing substantial resources to retrain existing employees or competing for costly external talent in a highly competitive job market.

UPSKILLING: A WIN-WIN FOR EMPLOYERS AND WORKERS

Several pressing talent challenges underscore the imperative for businesses to invest in retraining and upskilling: the scarcity of skilled professionals for new tech-centric roles, the significant resources required to attract external talent, the erosion of institutional knowledge due to staff turnover, diminishing competitive edge against AI-forward rivals, and escalating skills gaps that contribute to rising unemployment rates.

While the scale of disruption is daunting, it presents exciting upskilling opportunities through training programs tailored to build capabilities for new roles. Amazon is heavily investing in upskilling and reskilling initiatives to build employees' AI capabilities with their Machine Learning University.[5] Adobe's Digital Academy helped several employees transition to data-focused roles, and upskilled employees saw a salary increase of thirty to one hundred thousand dollars.[6] Deloitte has launched the Deloitte AI Academy, a program aimed at educating the next generation of AI professionals, as part of Deloitte's initiative to train up to ten thousand professionals in AI. Deloitte AI Academy has expanded its curriculum to include generative AI training that both practitioners and clients can tap into.[7]

These examples demonstrate that targeted reskilling delivers manifold benefits beyond filling open positions by:

- preserving valuable institutional knowledge.
- motivating and engaging employees by investing in their growth.
- showing commitment to workforce employability and mobility.
- allowing filling of critical skills gaps from within.
- saving costs of external recruitment and onboarding.

With adaptable vision, companies can cultivate the talent they need within existing teams. Reskilling helps workers remain relevant amid seismic technological shifts.

To successfully harness the power of reskilling, businesses can implement various strategic initiatives. These approaches ensure the workforce not only adapts to but also thrives in an evolving technological landscape. Here are some best practices for executing effective workforce upskilling:

- Conduct skills gap analysis to identify role-based deficits.
- Curate customized programs, blending technical and soft skills.
- Offer incentives like bonuses, promotions and certifications.

- Seek partnerships with online learning platforms and academia.
- Promote a culture of continuous learning and growth mindset.

MANAGING A MULTIDISCIPLINARY AI TEAM

At the heart of any successful AI project is a diverse, multidisciplinary team. Convening such a multifaceted group is truly an art form that, when done right, leads to remarkable innovation.

First and foremost, take time to understand the composition of your AI team. Recognize the distinct perspective each role brings—technical, design, or ethical. Foster open communication across disciplines through team meetings, open-door policies, and collaborative tools. Knowledge sharing is key, so organize cross-training sessions for team members to explain fundamentals of their expertise. Build an empowering toolkit with collaborative platforms like Jira, Slack and GitHub. Seek regular feedback to continuously improve operations. Above all, focus on building trust between team members—the cornerstone of successful collaboration.

Actively celebrate the diversity of thought and problem-solving approaches within your team. Provide growth opportunities through training and leadership roles to retain top talent. Establish efficient workflows between interdependent roles to maximize productivity. Don't underestimate the value of soft skills; emphasize communication, empathy and adaptability in addition to technical prowess.

Set clear boundaries between roles while encouraging peer mentorship opportunities. Align every decision with overarching business goals. Motivate your team through celebrating small wins, analyzing failures as lessons learned, and providing the right tools and resources. Like a conductor guiding an orchestra to perform a masterpiece, lead with vision, communication, and trust. When done right, your multidisciplinary AI team will make beautiful music!

Set clear, measurable objectives upfront to align the team toward a unified vision. Adopt agile methodologies to enhance flexibility amid the dynamic nature of AI work. Anticipate and resolve conflicts through open dialogue, leveraging them as opportunities for innovative solutions. Managing a diverse AI team requires understanding individual strengths, fostering collaboration, establishing efficient workflows, aligning with business goals, and building trust. Do this well, and you empower a dream team primed for groundbreaking innovation.

CASE STUDY: SUCCESS THROUGH A MULTIDISCIPLINARY AI TEAM

In large organizations, employees often find themselves navigating a labyrinth of separate systems for IT support, HR queries, finance, and facilities management. This fragmentation leads to decreased productivity, increased frustration, and higher operational costs. Moveworks aimed to streamline this convoluted process by creating an AI-driven platform that operates within Microsoft Teams, automating common tasks and workflows.

Moveworks developed an AI bot built on Microsoft Azure that integrates seamlessly with Microsoft Teams. The bot is designed to understand enterprise context and business jargon in over one hundred languages, making it accessible to a global workforce. It connects to existing enterprise systems, pulling relevant information, and forms directly into Teams, allowing employees to stay within a single platform and eliminating the need to switch between multiple apps and learn new workflows.

The triumph of Moveworks' AI-driven solution is a testament to the synergy of a multidisciplinary team. Machine learning and AI experts crafted over two hundred models to decode enterprise jargon while software developers integrated the bot seamlessly with platforms like Microsoft Teams and Azure. Enterprise systems specialists ensured compatibility with existing organizational systems, from HR to finance. Multilingual NLP experts expanded the bot's reach to over one hundred languages, and UX designers honed an intuitive user interface.

Orchestrating this diverse skill set, project managers kept the team aligned and on schedule, proving that a well-coordinated, multifaceted team is crucial for solving complex problems in the AI landscape. Since its launch four years ago, Moveworks has garnered a clientele that includes dozens of Fortune 500 companies. The platform has saved countless employee hours, significantly improved work efficiency, and reduced millions of dollars in IT support costs.[8]

The age of AI calls for perceptive, ethical leaders who can assemble diverse dream teams. With technology automating tasks, the human skills of creativity, empathy and purpose matter more than ever. By bridging technical expertise with business acumen, fostering trust, and upholding values, leaders can orchestrate AI across companies well. This prepares companies to harness AI as a force for innovation and progress, not just productivity.

CHAPTER 9
AI for Good: Developing AI Responsibly and Ethically

A LIFE-ALTERING DAY IN FARMINGTON HILLS

On a sunny afternoon in 2020, Robert Williams was arrested in front of his home and family in suburban Detroit. Unknown to Williams, a faulty facial recognition algorithm had wrongly matched his driver's license photo to security footage from a theft. Williams spent a traumatizing night in jail before being released with an apology. This cautionary tale ignited public scrutiny about the ethical implications of deploying AI technologies, like facial recognition, especially in sensitive domains like law enforcement.[1]

The narrative of Robert Williams is a stark reminder of the stakes at hand. As AI becomes more embedded in our lives and businesses, we have a collective responsibility to uphold human dignity and rights. Robust governance structures and oversight processes are crucial to align AI systems with company values and avoid pitfalls. Safety considerations, both cyber and operational, should be baked into AI development. Finally, historical, statistical and representation biases must be actively identified and mitigated.

ETHICAL AI PRINCIPLES AND PRACTICES

In the journey of AI integration, ethical considerations are paramount. They help businesses harness the power of AI responsibly, benefiting not just their bottom line but society at large. By embedding ethics into the very fabric of AI development and deployment, businesses not only mitigate risks but also position themselves as responsible and forward-thinking entities in the AI landscape. This approach ensures that as we sail further into the AI-driven future, we do so with a moral compass firmly in hand.

To address the ethical challenges, consider the following principles and practices:

- **Value Alignment:** Align the AI system with the core values and goals of your company, reflecting principles such as patient well-being in health care, where algorithms should prioritize medical needs above all else.
- **Transparency:** Ensure the AI system not only makes decisions but also explains how those decisions are reached, particularly in sectors like financial services where decisions significantly impact lives.
- **Fairness:** Work actively to mitigate biases in your AI systems. This involves everything from diverse data collection to algorithmic fairness checks.
- **Accountability:** Establish processes to audit AI systems. If an AI system makes a mistake, you should have a clear method for identifying what went wrong and how to prevent it in the future.
- **Privacy:** Implement robust data protection measures to safeguard sensitive information, recognizing the importance of data privacy in AI operations.

- **Safety and Security:** Ensure the AI system is safe to use and secure from external threats, such as encrypting patient data and rigorously testing diagnostic algorithms in health care.

SHARED PATHS: EVERYONE'S ROLE IN ETHICAL AI

Ethical AI isn't just the responsibility of the tech team or the C-suite; this collective endeavor involves various stakeholders. As Potter Stewart wisely said, "Ethics is knowing the difference between what you have a right to do and what is right to do."[2]

Employees are often the first to interact with AI systems and can identify potential ethical pitfalls or biases. To empower employees, businesses should provide comprehensive training on AI ethics. Additionally, companies should establish open channels for employees to report ethical concerns or system flaws, making them active participants in the ethical oversight of AI.

Customers, the end-users, are the ultimate test of an AI system's ethical alignment. Transparency is key here. Businesses should clearly explain how AI impacts the services or products customers receive. Informed consent is crucial when using customer data, and you should have avenues for customers to dispute AI decisions that impact them.

Shareholders have a vested interest in the company's long-term success and sustainability. Regular updates on ethical AI practices and audits can keep them informed. Ethical AI is not just a cost but an investment in long-term value, a point that should be emphasized to shareholders. Governance structures, such as specialized committees, can involve them in ethical oversight.

GOVERNANCE AND OVERSIGHT

As AI becomes increasingly integrated into business operations, the need for robust governance and oversight is vital. Let's explore the role of the board and executives, delve into frameworks for governance, and discuss the importance of regulatory compliance.

THE ROLE OF THE BOARD AND EXECUTIVES

The board and executives hold the reins of AI governance. They are responsible for developing policies for acceptable AI use that align with the company's values and goals. Oversight processes and committees should be established to govern AI systems effectively. These committees can conduct regular impact assessments to identify risks and ensure the AI systems are both ethical and effective.

Consider the example of a financial services company that faced challenges with AI-driven risk assessment algorithms. The board and executives took proactive steps to establish an AI Ethics Committee. This committee conducted regular audits and impact assessments, leading to the identification and mitigation of biases in their algorithms. As a result, the company not only enhanced its ethical standing but also improved customer trust and compliance.

FRAMEWORKS FOR GOVERNANCE

One of the most respected frameworks for AI governance is the AI Ethics Guidelines by the Institute of Electrical and Electronics Engineers (IEEE). This framework offers a comprehensive set of guidelines that cover everything from value alignment and transparency to accountability and safety.[3] Companies can adapt these guidelines to fit their specific needs and industry requirements. Compliance with laws and regulations is nonnegotiable. In the context of AI, this often involves data protection laws such as the General Data Protection Regulation (GDPR) in Europe and the California Consumer Privacy Act (CCPA) in the United States. These laws mandate stringent data protection measures and offer guidelines on how personal data should be handled.

- **Develop Policies:** Create comprehensive policies that outline acceptable AI use, ensuring alignment with existing laws and ethical guidelines.
- **Oversight Processes:** Establish committees or designate responsible individuals to oversee AI systems.
- **Impact Assessments:** Conduct assessments regularly to identify risks, including ethical risks associated with AI use.
- **Responsible Scale-up:** Build scalable governance mechanisms for your growing AI initiatives. This will enable responsible scale-up of AI usage without compromising on ethical or legal obligations.

NIST'S RISK MANAGEMENT FRAMEWORK

The National Institute of Standards and Technology (NIST) has developed the AI Risk Management Framework (AI

RMF) to give a strong foundation for responsible governance. Released in January 2023, the AI RMF offers a set of guidelines to help organizations manage the risks associated with AI systems throughout their lifecycle.[4]

The core objective of the AI RMF is to enhance the integration of trustworthiness considerations into every stage of AI development, use, and evaluation. This framework provides a structured approach for identifying potential risks, assessing their severity, and implementing mitigation strategies. By following the AI RMF, organizations can expect several benefits, including:

- **Comprehensive Risk Management:** Organizations gain a deeper understanding of the relationships between trustworthiness characteristics, socio-technical considerations, and potential AI risks.
- **Flexibility and Measurability:** The AI RMF provides a flexible yet structured and measurable process for addressing AI risks.
- **Responsible Innovation:** By following the framework, organizations can maximize the positive impact of AI technologies while minimizing potential negative consequences for individuals and society.

SAFETY FIRST

Nelson Mandela once said, "Safety and security don't just happen. They are the result of collective consensus and public investment."[5] In the context of AI, this means safety should be a cornerstone, not an afterthought.

AI AND DATA SECURITY

AI has a paradoxical relationship with data security; it can both compromise and enhance it. On one hand, sophisticated AI algorithms can be used in cyber-attacks, making traditional security measures obsolete. On the other hand, AI can be a powerful ally in identifying vulnerabilities and thwarting cyber threats real time.

Consider a cybersecurity firm that uses AI to monitor network traffic. While the AI system can identify and neutralize threats much faster than human analysts, it's also a potential target for hackers who could manipulate it to gain unauthorized access. Therefore, setting safety as a key criterion in the development phase is crucial. Techniques like uncertainty modeling can help the system recognize when it's facing a situation it hasn't been trained for, triggering additional security protocols.

OPERATIONAL SAFETY

Operational safety takes center stage when AI is applied in sectors like manufacturing or logistics. In these settings, AI systems must prioritize both security and operational safety to prevent accidents and maintain smooth workflows.

In a manufacturing plant, while robots can significantly improve efficiency, any malfunction could lead to serious safety hazards. A risk management framework should be in place to assess and mitigate such risks. Safe exploration techniques can be used to test new operational strategies in a simulated environment before deploying them in the real world. Human oversight is essential to intervene in case the AI system behaves unexpectedly.

MITIGATION AND OVERSIGHT STRATEGIES

As we embrace the immense potential of AI, we must also implement robust strategies to mitigate risks and maintain human oversight. Responsible AI development demands a proactive approach to address safety concerns and ethical implications from the outset. The following mitigation and oversight measures are crucial to ensure AI remains a positive force that augments and empowers humanity:

- **Set Safety Criteria:** Make safety nonnegotiable during the development phase of AI.
- **Uncertainty Modeling:** Use techniques that allow the AI system to recognize and appropriately respond to unfamiliar situations.
- **Risk Management Framework:** Develop a framework that identifies, assesses, and mitigates risks with AI.
- **Human Oversight:** Always have a mechanism for human intervention to correct or override AI decisions when necessary.

UNDERSTANDING BIAS IN AI

Bias can be defined as a systematic error that leads to unfair or unequal outcomes. In AI, bias often creeps in through the data used to train models or the design of the algorithms themselves. There are various sources of bias to consider:

HISTORICAL BIAS

Data can reflect societal biases, such as gender or racial discrimination. Employment practices, particularly in STEM fields, continue to exhibit bias, with women holding less than a quarter of technical roles as of 2020 as reported by Deloitte.[6]

Amazon's AI recruitment tool we discussed in chapter 6 exemplified this issue. This tool, after analyzing historical hiring data, inadvertently learned to favor male candidates, leading to a systemic undervaluation of female applicants' résumés.[7] Despite efforts to correct this, the ingrained bias led to the program's discontinuation in 2017, highlighting the challenges of eliminating deep-rooted biases in AI-driven systems.

STATISTICAL BIAS

Statistical bias in AI occurs when the data used to train the model is not representative of the population it will serve.

One prominent example of bias in AI is found in facial recognition algorithms, which despite high overall accuracy rates (over 90 percent), show significant performance disparities across demographic groups. Research reveals these algorithms consistently perform poorly for certain groups, particularly Black females, and those aged eighteen to thirty. Evaluations found error rates up to 34 percent higher for darker-skinned females compared to lighter-skinned males, highlighting how underrepresentation in training data leads to statistical bias and inequities in real-world AI applications.[8]

Another example is the COMPAS algorithm used in US court systems to predict the likelihood that a defendant would become a recidivist. Due to the data that was used, the model predicted twice as many false positives for recidivism for black offenders as white offenders.[9]

REPRESENTATION BIAS

Representation bias happens when certain groups are underrepresented in the data, leading to skewed outcomes. This bias can impact decision-making in AI by leading to skewed outcomes.

AI algorithms can inadvertently perpetuate racial biases. For example, a health-care predictive model widely used in the US was found to be biased against Black patients. This model, designed to identify individuals needing extra medical care based on health-care costs, inadvertently assigned lower risk scores to Black patients compared to similar white patients. This discrepancy arose because the model used health-care costs as a proxy for medical need, failing to account for racial differences in how patients access and pay for care. Consequently, Black patients, despite having significant health issues, were often deemed less in need of additional care due to their lower health-care expenditures, thereby propagating unequal treatment.[10] This underscores the importance of evaluating algorithms for prejudices and designing them to promote fairness across racial groups.

Bias	What It Means	Example
Historical Bias	Reflects societal biases like gender or racial discrimination.	Amazon's AI recruitment tool favored male candidates.
Representation Bias	Certain groups underrepresented in data.	Healthcare predictive model assigned lower risk scores to black patients.
Statistical Bias	Data not representative of the population served.	COMPAS algorithm's false positives for recidivism among black offenders.

Table 6: Types of Biases in AI

MITIGATING BIAS IN AI SYSTEMS

Mitigating bias in AI is an ongoing commitment that requires a structured approach. The framework below outlines key steps and strategies for reducing bias throughout the AI lifecycle—from data collection to model deployment and beyond.

This framework emphasizes a proactive and comprehensive approach to bias mitigation, integrating technical adjustments with ethical considerations. By following these steps, organizations can work toward ensuring their AI systems are fair, equitable, and inclusive, thereby contributing positively to society and upholding ethical AI standards.

Bias Awareness and Identification: Recognize various forms of bias, including historical, statistical, and representation biases. Evaluate the data for diversity and representativeness, considering the context in which the AI system will operate.

Data Preparation and Diversification: Include data that encompasses a broad spectrum of demographics, behaviors, and scenarios to counteract historical and representation biases. Apply techniques to balance the data, such as oversampling underrepresented groups or under sampling overrepresented ones.

Algorithm Design with Built-In Fairness: Develop algorithms that account for ethical AI principles and fairness from the outset, potentially modifying the weighting of features to diminish the influence of biased data points. Incorporate tools like Fairness Indicators to continuously evaluate the model's predictions across different groups during development and after deployment. Perform ongoing bias assessments as new data is integrated to detect and mitigate any emerging discriminatory outcomes.

Continuous Refinement: Incorporate feedback from audits and testing to refine the model, adjusting datasets and algorithms as necessary to enhance fairness. Continually update AI systems with more representative data and retrain models to adapt to evolving contexts and mitigate emerging biases.

TRANSPARENT AI: ILLUMINATING THE BLACK BOX

A key tenet of developing AI responsibly is making sure it's transparent. Transparent AI, often referred to as explainable AI (XAI), is the practice of designing artificial intelligence systems with a layer of interpretability, allowing humans to understand and trust the decisions made by AI. This means creating AI that can explain its reasoning, detail its limitations, and describe its decision-making process in a way that is accessible to the average person.

Transparent AI is pivotal in fostering ethical AI practices and robust governance frameworks. It engenders trust among stakeholders by demystifying the decision-making process of AI, particularly in sectors where the implications are profound, such as health care and law enforcement.

In the context of regulatory frameworks, the European Union's Artificial Intelligence Act (AI Act) is a significant development. Passed in 2024, the AI Act is the world's first comprehensive legal framework on AI. It categorizes AI applications into three risk categories and sets out specific requirements for each. High-risk AI applications, such as those used in medical devices or critical infrastructure, face stringent requirements, including the use of high-quality data and providing clear information to users. The Act also bans certain AI uses deemed to pose an unacceptable risk. This legislation underscores the importance of transparency in AI as it mandates that AI systems respect fundamental rights, safety, and ethical principles.[11] The AI Act has set a global precedent, nudging the future of AI in a human-centric direction where humans are in control of the technology. This aligns with the principles of transparent AI, further emphasizing the need for AI systems to be understandable and trustworthy.

Transparent AI incorporates techniques that make the inner workings of AI algorithms visible and understandable. This can be achieved through various methods, such as:

- **Model Interpretability:** Using simpler models that are inherently more interpretable, such as decision trees, or applying techniques to complex models to extract decision rules.
- **Feature Visualization:** Highlighting which features the AI considers most important when making a decision.
- **Local Explanations:** Providing specific explanations for individual decisions.
- **Global Explanations:** Offering an overview of how the model behaves across a wide range of cases.

Having worked closely with EazyML, I can attest to their pioneering role in transparent AI. They provide a user-friendly platform that enables users to build, deploy, and interpret machine learning models without requiring deep technical expertise. By providing explanations in plain language, EazyML makes it possible for nonexperts to understand and trust the outputs of AI systems. This approach not only democratizes AI but also aligns with the growing demand for ethical AI practices, where understanding and trust are paramount.

Other companies such as IBM, Google, and Microsoft are providing various degrees of transparent AI. IBM offers open-source tools like AI Fairness 360 and AI

Explainability 360 to detect bias and explain AI decisions, democratizing transparent AI practices.[12] Google sets industry benchmarks with its AI principles and tools like TensorFlow Extended, which includes model understanding components.[13] Microsoft integrates transparency directly into Azure, offering interpretability features that facilitate the deployment of transparent AI at scale.[14]

AI FOR GOOD

Exploring the potential of artificial intelligence to serve as a catalyst for positive change goes beyond mere technical achievements. The essential question is not whether AI can think but how we can harness its capabilities for the greater good.

The potential of AI to foster advancements in critical areas such as health care, sustainability, and social justice is immense. In health care, AI's precision in analyzing medical imagery heralds a new era of early disease detection, with a significant impact on conditions like cancer. The optimization of resources and processes through AI not only enhances sustainability but also brings efficiency to manufacturing, reducing environmental footprints.

Moreover, AI's ability to sift through extensive datasets unveils patterns in human rights violations, offering a new lens to address issues like trafficking and discrimination. This technology democratizes accessibility, breaking barriers related to socioeconomic status, ethnicity, or geography. For instance, AI-driven telemedicine extends health-care services to remote regions, ensuring vital services reach those in underserved communities. By harnessing AI, we open avenues to not only refine existing systems but also to craft innovative solutions that address complex societal challenges, making the concept of "AI for Good" a tangible reality.

SECTION III

TRANSFORMATION: THE ART OF POSSIBILITY

CHAPTER 10

Moneyball: How AI Is Revolutionizing Finance

Last year, I had the privilege of leading a groundbreaking project for one of our clients, a major player in the global computer hardware industry. This company was facing significant challenges with its accounts receivable process. With over a thousand customers to manage, their three-hundred-member billing department was deeply entrenched in a manual, error-ridden invoicing system. This inefficient process led to 10 percent of invoices needing rework due to inaccuracies, which in turn doubled the payment receipt timeline from the expected sixty days to an extended one hundred and twenty days.

Recognizing the need for a transformative solution, we introduced an AI-powered system designed to streamline and automate the invoice-to-cash workflow. The new AI system dramatically reduced the error rate to 2 percent, liberated nearly half of the team's time previously dedicated to manual processing, and restored operational efficiency, achieving the target of sixty days sales outstanding. Not only did it enhance cash flow by over twenty million dollars, but it also yielded annual operational cost savings of seven million dollars.

This case study is an example of the transformative power of AI in finance.

Think of a future where the monotonous tasks in finance—from manual invoice processing to the continuous monitoring of accounts payable—become things of the past. This future is on the horizon, thanks to the advent of AI in finance. The potential of this shift is monumental. According to research by McKinsey, AI could automate more than 40 percent of tasks within the financial domain, ushering in an unprecedented level of efficiency and accuracy.[1] AI is reimagining every facet of finance—from budgeting practices to regulatory compliance—heralding a future where strategic initiatives eclipse the mundane, manual efforts that once dominated the field.

ACCOUNTS PAYABLE AUTOMATION

Consider the benefits of an AI assistant in your AP department that acts like a "virtual accountant." This intelligent bot could monitor your shared inbox and respond to most vendor inquiries autonomously. Leveraging natural language processing and generative AI, it would converse with vendors about approvals, payments, and other workflow activities. The bot could send documentation on request, improving vendor satisfaction. The impact would be transformative. Response times would be lightning fast with twenty-four-seven availability. Bots can handle fluctuating inbox volume and reduce your team's workload. Properly trained machine learning engines underlying the bot would make AP highly accurate. AI would also provide real-time visibility into communications. This gives AP managers data-driven insights to optimize processes and key metrics. The rise of intelligent bots in AP is coming.

A major corporation, which remains anonymous at the request of a finance executive there, faced challenges with inefficient, manual accounts payable processes that caused frustration among vendors. The small AP team was inundated with vendor inquiries flooding their shared inbox, leading to slow responses and error-prone payments. To transform their accounts payable operations, the company implemented an AI-powered intelligent virtual assistant that automated vendor communications and document management.

The AP bot could handle the high-volume inbox autonomously, leveraging NLP capabilities. It rapidly responded to inquiries, accurately resolved issues, and provided documentation on-demand around the clock. Within a few months, vendor satisfaction increased dramatically, while inquiry response times declined from forty-eight hours to less than one hour. The accounts payable team's productivity improved by over 30 percent as staff could focus on value-added tasks instead of reading emails. Payment errors decreased by 17 percent. Real-time analytics provided data-driven visibility, enabling the company to optimize processes further.

INVOICE PROCESSING AND MANAGEMENT

Invoice processing is one of the most tedious and time-consuming tasks in finance. Manual errors can happen during invoice coding, general ledger (GL) mapping or payment detail verification. With autopilot data entry, AI can extract key details from invoices and input them without any human review needed. It can also automatically classify costs and route invoices to the right people for approval. Even better, AI can match invoices to purchase orders and handle exceptions. Manual invoice processing is only about 60-70 percent accurate on average. But AI can improve accuracy to 98 percent by eliminating human errors.[2]

AI provides business insights by detecting spending patterns and opportunities for cost savings. It delivers real-time operational analytics to identify bottlenecks and issues per user, vendor, and entity. With millions of data points indexed, AI can even recommend prescriptive actions to optimize processes.

Vic.ai is an AI platform designed to help companies streamline invoice processing. The platform uses AI to run the accounting process from beginning to end, including invoice entry, classification, matching, and approvals. With AI technology, it can extract the number on an invoice, understand and classify the cost, send it to the right person for approval, or automatically approve it if it meets all the criteria. The platform can also provide predictive reasoning to make decisions about an invoice after seconds of ingestion, helping to save time and reduce errors. Some of Vic.ai's clients were able to automate invoice processing and reduce time spent on processing by as high as 80 percent.[3]

FINANCIAL REPORTING

For many finance teams, financial reporting is filled with manual data uploads, endless validations, rushed closes, and delayed insights. Rather than manually moving and checking datasets, AI handles the task seamlessly with high accuracy. Finance teams could save a lot of time and trust the data used for reporting. Next, AI establishes a crystal-clear baseline of historical performance by crunching volumes of financial data. This helps the teams spot trends and patterns over time to set targets for the future. When it comes to the financial close, AI can optimize the entire process. Collecting, summarizing, and analyzing data is fast. Close cycles could shorten from weeks to days. Finally, AI generates on-demand reports and analysis conveying timely and accurate financial results. This empowers finance teams to make swift, data-driven decisions about the business while ensuring regulatory compliance.

In a practical application of this technology, Hershey has integrated Workiva's AI-powered platform, overhauling its approach to financial reporting. With the AI technology, they streamlined the creation of annual and quarterly SEC filings and expanded its capabilities to encompass a wider array of regulatory documents. The AI feature that stands out the most is the ability to link a single data entry to multiple reports, which then automatically synchronizes updates across all occurrences. This not only ensures data consistency and accuracy but also eliminates extensive hours of manual revision. Hershey's reporting team has benefited from efficient management of last-minute changes and the capacity for collaborative real-time document editing. This technological leap has shifted Hershey's focus from the administrative tasks of reporting toward more strategic financial analysis and decision-making.[4]

FINANCIAL PLANNING AND ANALYSIS

A multitude of challenges often encumber financial planning and analysis (FP&A) teams, impeding their efficiency and effectiveness. The sheer volume of data available to FP&A professionals can be overwhelming, making it difficult to extract actionable insights. This data deluge coupled with an excess of manual, time-consuming processes, diverts attention from strategic analysis to mundane data manipulation.

Moreover, traditional reporting methods are rigid and fail to keep pace with the rapid changes in the business environment, leading to outdated insights. Disparate systems and data silos further exacerbate the issue, creating barriers to a holistic view of analytics. Additionally, the complexity of modeling various business scenarios and projections is amplified by the inherent uncertainty and multitude of variables at play.

By automating the integration of data from various systems, AI facilitates the creation of cohesive datasets, thereby streamlining analysis. AI's advanced forecasting algorithms can swiftly process and analyze numerous variables and potential what-if scenarios, yielding more accurate and nuanced projections. This capability enables dynamic, customizable, and conversational insights, empowering finance leaders with on-demand analytics accessible through simple language queries. AI significantly enhances scenario planning and risk management by employing probability modeling and in-depth scenario analysis, allowing FP&A teams to anticipate various future scenarios more effectively. Additionally, AI-driven anomaly detection is instrumental in uncovering irregularities, potential threats, and nascent trends that might elude manual analysis. This detection safeguards data accuracy and minimizes risks, providing FP&A teams with the foresight and agility needed for strategic, informed decision-making.

Leading corporations are already harnessing the power of AI to redefine their FP&A functions. PepsiCo, for instance, has integrated AI to automate forecasting and financial reporting processes, which has dramatically reduced the time required to close books from days to mere hours, simultaneously enhancing the accuracy of their financial data. Their AI assistant allows for self-service analytics, enabling users to ask questions and receive insights in natural language.[5]

AI IN RISK MANAGEMENT AND COMPLIANCE

Managing financial risk is a complex challenge, traditionally burdened with manual tasks like data collection, analysis, and reporting. The advent of AI is significantly streamlining these processes. In financial data analysis, transparent AI models are proving to be invaluable. These models excel at dissecting complex financial data and identifying potential risks with remarkable accuracy.

Their transparency is key as it allows financial institutions to understand what factors the AI model has considered. Finance professionals can modify the factors if appropriate and have faith in AI's risk assessments, thereby making more informed decisions. This clarity is crucial not only for internal decision-making but also for satisfying regulatory and auditor requirements.

AI is helping insurance companies better understand customer risk, leading to accurate premiums. Through AI-driven probability modeling and scenario analysis, these companies can now anticipate a wide range of potential outcomes with greater precision. Moreover, AI-powered anomaly detection algorithms play a pivotal role in uncovering deviations and emerging trends, enabling insurers to identify risks that may not be evident through traditional analysis methods.[6]

On the compliance front, AI is equally transformative. It automates workflows, identifies risks early, and provides insights that might elude human analysis. This capability is particularly beneficial in adhering to complex and evolving regulations. For instance, AI can proactively detect compliance violations, allowing companies to address issues before they escalate. Additionally, AI aids in managing regulatory changes by monitoring and analyzing regulatory frameworks, ensuring that financial operations remain compliant.

A practical example of AI's role in compliance is its ability to detect potential regulatory violations proactively. Transparent machine learning models, such as those provided by EazyML, can explain the rationale behind each prediction, allowing experts to scrutinize and respond to alerts. In cases of suspected bias leading to false alerts, the MLOps team can intervene, investigate, and retrain the model to eliminate biases, thereby maintaining compliance.

FRAUD DETECTION AND PREVENTION

Recently, I had the opportunity to catch up with Manish Gupta to delve into American Express's groundbreaking journey of integrating AI into its fraud detection systems back in 2018. Under Gupta's leadership, the team tackled the inherent limitations of traditional systems in pinpointing intricate fraudulent patterns by embracing deep learning models, particularly focusing on LSTM (Long Short-Term Memory) models known for their adeptness in analyzing sequential data. This venture required overcoming substantial hurdles, including aligning the AI technologies with the stringent regulations of the financial sector and ensuring the system's capability for real-time processing of the vast volume of transactions. A pivotal aspect of this initiative was the strategic partnership with Nvidia, which provided the specialized hardware and software essential for adhering to the demanding service level agreements (SLAs) associated with real-time fraud detection.

By 2021, the AI-enhanced fraud detection initiative had markedly improved the accuracy of identifying fraudulent activities and set a new benchmark in the financial industry for the application of technology in enhancing security protocols. The integration of LSTM models as key components within AmEx's existing fraud detection framework underscored the significant advantages of amalgamating AI innovations with conventional systems. Despite initial challenges such as regulatory constraints and the necessity for specific infrastructural adjustments, the concerted efforts of Gupta's team and the judicious application of technology culminated in the development of an advanced system capable of real-time processing and precise fraud detection. This conversation with Gupta highlighted the transformative potential of AI in bolstering the fraud detection capabilities of financial institutions, signaling an optimistic future for the continued evolution of AI in this critical area.

With the rise of digital transactions, fraudsters are exploiting new technologies to orchestrate sophisticated scams—from phishing to unauthorized transfers—broadening the fraud spectrum and undermining trust. Traditional fraud prevention methods, based on static rules and manual checks, struggle to counter these evolving tactics. This scenario underscores an urgent need for financial institutions to embrace more advanced, intelligent systems that can analyze extensive data, detect anomalies, and swiftly intervene, enhancing security and customer confidence.

This is where artificial intelligence delivers transformative capabilities. One of the standout innovations in fraud prevention is behavioral biometrics, which analyzes unique human behavior patterns, such as typing rhythm, mouse movements, and touchscreen interactions. This approach goes beyond merely knowing a user's password; it delves into understanding their behavior. Coupled with continuous authentication, systems no longer just verify users at login. They monitor user behavior throughout their session, tracking actions such as typing speed and style, navigation patterns, time spent on different tasks, and even pressure applied on touchscreens. This continuous monitoring creates a dynamic profile of normal user behavior, allowing the system to detect deviations that may indicate fraudulent activities. By offering a persistent layer of security, continuous authentication provides a robust defense against unauthorized activities, ensuring that any suspicious behavior triggers immediate alerts and interventions to protect both the user and the institution.

Machine learning, a key component of AI, is crucial for advancements in fraud detection. By analyzing vast datasets in real time, AI algorithms can quickly identify patterns and anomalies indicative of fraud. This enables financial institutions to detect and prevent fraudulent activities with greater accuracy. Additionally, transparent AI systems provide clear evidence for suspicious activities, aiding investigators and promoting trust among customers. This transparency strengthens the security of financial institutions.

Another challenge in the financial industry is verifying the identity of entities, especially when data is fragmented or inconsistent. Entity resolution addresses this by determining the real-world identity of entities from various data sources. Moreover, network analysis maps out transaction networks, spotlighting suspicious activities or connections.

The introduction of decision logic uses machine learning to make real-time decisions on transactions. The concept of dynamic friction adds extra verification steps for suspicious activities while ensuring a seamless experience for genuine users. Link Analysis further identifies connections between seemingly unrelated activities, revealing coordinated fraud attempts. The use of device intelligence delves deep into studying devices used for transactions to flag potential risks. Cross-Channel Analysis monitors user activity across various platforms, ensuring comprehensive fraud detection. Biometrics has also made a significant mark in the fraud detection arena. Beyond just fingerprints, the industry now uses voice, facial, and behavioral biometrics for authentication. Conversational AI analyzes user interactions and conversations to detect fraud attempts by monitoring for unusual language patterns, inconsistencies in responses, and deviations from expected conversational behaviors, which may indicate fraudulent activities.

Lastly, transaction monitoring is of paramount importance. By observing all transactions in real time, suspicious activities are quickly flagged. Systems today employ Adaptive Analytics, continuously refining their fraud detection capabilities based on new data. The creation of global intelligent profiles offers a holistic view, tracking behavior over time and across various channels.

PayPal uses machine learning algorithms to analyze millions of transactions in real time to detect and prevent fraud. The system can identify patterns and anomalies that indicate fraudulent activity and take immediate action to stop it.[7] Capital One uses AI-powered fraud detection tools to monitor customer accounts and transactions for suspicious activity. The system can analyze data from multiple sources, including social media and public records, to identify potential fraudsters.[8] Mastercard uses AI to analyze transaction data and identify patterns that indicate fraudulent activity. The system can also detect anomalies in user behavior, such as sudden changes in spending patterns or unusual locations, to prevent fraud.[9]

While AI is leveraged for fraud detection, it's imperative that customer experience is not compromised. For instance, if a genuine credit card transaction is declined due to a false positive by machine learning algorithms, customer experience is compromised. Potential business may be lost.

WEALTH MANAGEMENT

AI is also making significant inroads into wealth management. Sophisticated algorithms analyze market trends, investment options, and individual financial goals to offer tailored wealth management solutions. BlackRock, one of the world's largest asset managers, uses AI to optimize asset allocation, tax strategies, and even estate planning. Their Aladdin platform employs machine learning to analyze market risks and opportunities, offering a comprehensive wealth management service.[10]

AI-driven platforms provide personalized investment advice based on an individual's financial goals and risk tolerance. Betterment's robo-advisor uses AI to provide personalized investment advice. By analyzing an individual's financial goals, risk tolerance, and history of interactions, Betterment offers tailored investment portfolios, making wealth management accessible to the masses.[11]

Similar to Betterment, Wealthfront uses AI to offer personalized financial planning. It considers various factors like age, financial goals, and risk appetite to suggest the most suitable investment options.[12]

TRANSFORMING PROCUREMENT WITH AI

The rise of AI platforms like Globality is disrupting traditional procurement processes, empowering companies to find the best-fit suppliers based on merit rather than defaulting to familiar incumbents. Joel Hyatt, CEO of Globality, shared a compelling example of how their AI technology transformed a global pharmaceutical company's digital marketing efforts.

Dissatisfied with their existing agency's performance, the pharma giant turned to Globality's AI-driven matchmaking platform. By evaluating a diverse pool of suppliers against the company's specific needs, Globality's algorithms surfaced a surprising contender—a small, local boutique firm just 1.5 miles away that the pharma company had previously been unaware of. Despite its modest size, this boutique agency's innovative strategies and creative concepts impressed during the AI-facilitated evaluation process, leading the pharmaceutical company to award them the business over larger, more established incumbents.

The results were remarkable. The boutique agency's digital marketing campaign generated four times higher return on investment compared to previous campaigns while simultaneously reducing costs by 50 percent. Through objective, AI-driven analysis, the pharmaceutical company identified a superior supplier that unlocked dramatic performance improvements.

For the boutique firm, Globality's platform provided crucial access, enabling them to compete for and win a major client that would have been out of reach through traditional procurement channels. AI empowered a meritocratic process focused on quality and performance rather than relying on existing relationships or preconceptions about a supplier's size or reputation.

Procurement teams today face immense pressure to deliver greater value and efficiency. Growing business complexity, exploding supplier networks, massive amounts of data, and the need for agility are pushing traditional procurement processes to their limits. We've all been there—the daunting task of responding to requests for proposals (RFPs). Think of the possibility where AI can swiftly analyze RFPs, understand intricate requirements, and even draft preliminary responses. This is where artificial intelligence delivers transformative capabilities.

Managing vendors can sometimes feel like herding cats. But AI comes to the rescue by meticulously tracking vendor performance, analyzing feedback, and ensuring strict contract compliance. Companies like Coupa, which provides software for business spend management (BSM), are harnessing AI to keep a keen eye on vendor performance, ensuring businesses always get the best out of their partnerships.[13] Amazon's procurement system harnesses the power of AI to analyze vendor performance meticulously, ensuring they always get the best value for their money.[14]

Contract management is another area where AI is proving to be highly valuable. AI enables businesses to extract, review, and analyze critical contract information with unparalleled speed and accuracy. With this capability, companies are able to identify and deliver obligations before deadlines, avoiding costly penalties and customer churn. Some standout features of advanced contract AI tools include:

- **Optical Character Recognition (OCR):** Transforming scanned documents into editable and searchable data.
- **Clause Extraction:** Pinpointing and extracting essential data and clauses from contracts.
- **Contract Classification:** Categorizing contracts and tracing their origins.
- **Unique Semantic Search:** Offering a deep, meaningful search within contracts.
- **Clause Comparison:** Comparing clauses across contracts and identifying deviations.

Multiple AI techniques are used to help procurement. Generative AI and the\ underlying LLMs are used to analyze contracts and RFPs as well as to extract key terms and requirements. This can help procurement teams quickly identify relevant information and make informed decisions. Machine learning algorithms can analyze large datasets to identify patterns and make predictions. In procurement, machine learning or generative AI can be used to analyze supplier performance, spend patterns, and inventory levels to provide insights and recommendations for procurement strategies.

Predictive analytics using machine learning analyzes data and makes predictions about future events, demand, potential supply chain disruptions, and inventory levels. AI-powered chatbots can assist procurement teams with supplier onboarding, purchase order creation, and invoice processing.

ENHANCING FINANCIAL STRATEGY AND CAPITAL ALLOCATION

As AI continues to advance, it is poised to become an invaluable tool for chief financial officers (CFOs) in optimizing financial strategy, capital allocation, and investment analysis. By leveraging AI's capabilities in data processing, pattern recognition, and predictive modeling, CFOs can gain deeper insights into their organizations' performance, market trends, and potential risks and opportunities.

BUSINESS MODEL ASSESSMENT AND OPTIMIZATION

AI can assist CFOs in evaluating the effectiveness of their company's business model by analyzing vast amounts of historical data, industry trends, and consumer behavior patterns. Through machine learning algorithms, AI can identify inefficiencies, pinpoint areas for improvement, and recommend data-driven strategies to enhance revenue streams, reduce costs, and optimize operational processes.

CAPITAL AND INVESTMENT ALLOCATION

One of the critical responsibilities of a CFO is to allocate capital effectively across various initiatives and investments. AI can aid in this decision-making process by analyzing a multitude of factors, such as market conditions, competitive landscapes, risk profiles, and projected returns. By leveraging AI agents and AI's predictive capabilities, CFOs can make what-if analyses and more informed decisions about where to allocate resources, ensuring capital is deployed in areas that yield the highest potential returns while minimizing risks.

RETURN ON INTERNAL INVESTMENTS

Evaluating the performance and return on internal investments is a challenging task, often involving complex calculations, projections, and assumptions. AI can streamline this process by continuously monitoring key performance indicators (KPIs), analyzing project data, micro, and macroeconomic trends, and forecasting future outcomes based on historical trends and external factors. This empowers CFOs to make data-driven decisions regarding the continuation, restructuring, or termination of internal investments, optimizing resource allocation and maximizing returns.

SCENARIO PLANNING AND STRESS TESTING

For businesses, scenario planning and stress testing have become crucial for mitigating risks and ensuring financial resilience. AI can simulate various scenarios by integrating multiple data sources, such as economic indicators, industry trends, and competitive dynamics. This enables CFOs to proactively identify potential threats and opportunities as well as develop contingency plans to navigate different scenarios effectively.

Functional Area	AI Opportunity
Accounts Payable Automation	• Virtual assistant handles vendor communications • Automated document management
Invoice Processing	• Data extraction and entry • Automatic routing and approvals • Identify spending pattern
Financial Reporting	• Automate data loading and validation • Accelerate financial close cycle • Generate on-demand reports
Financial Planning & Analysis	• Integrate data from various systems • Insights and advanced forecasting • Scenario modeling and risk assessment
Risk Management & Compliance	• Early detection of risks and violations
Fraud Detection & Prevention	• Behavioral biometrics • Real-time sophisticated fraud detection and analysis
Procurement	• Find optimal suppliers, simplify RFP process • Track vendor performance
Financial Strategy	• Business model assessment and optimization • Capital and investment allocation • ROI investments • Scenario planning and stress testing

Table 7: AI Opportunities in Finance

The integration of AI across various financial functions marks a significant leap toward efficiency, accuracy, and strategic decision-making. From transforming accounts payable with virtual assistants to leveraging advanced algorithms for fraud detection, AI is not just automating tasks. It's redefining the landscape of finance. As we continue to harness these technologies, finance stands on the brink of an unprecedented era of innovation, where manual processes give way to intelligent, data-driven insights, ultimately driving greater value for businesses and their stakeholders.

CHAPTER 11

The Personal Touch: Crafting AI-Driven Customer Experiences

For years, I dealt firsthand with the myriad of challenges plaguing customer experience (CX) as I had large CX teams in my organizations. The issues ranged from error-prone call routing and a disorganized knowledge base that stretched the learning curve for customer specialists to inadequate workforce planning, not to mention the lack of visibility into customer insights.

The overarching goal was clear. Reduce customer friction and enhance the customer journey at every touchpoint. We began by reimagining our approach to customer interactions, shifting from a traditional inside-out perspective to a customer-to-inside the organization approach. This paradigm shift was crucial. It meant that every process, every decision, was now made with the customer's experience at the forefront.

Automation and AI played a pivotal role in this transformation. We employed advanced analytics to gain a deeper understanding of customer behaviors and preferences. This data-driven approach enabled us to anticipate customer needs and proactively address potential issues before they escalated into problems. We streamlined our processes from call routing to issue resolution for efficiency. Our customer specialists were now equipped with tools that provided them with quick access to relevant information, reducing resolution times and improving the accuracy of their responses.

The impact of these changes was profound. We achieved a remarkable 90+ percent customer satisfaction rate, peaking at 95 percent. This was not just a testament to the effectiveness of automation and AI but also to the power of a customer-centric approach.

Can you see the nirvana where every customer feels truly understood, valued, and catered to as an individual? This extraordinary level of personalized experience is becoming a reality through the transformative power of AI. AI is altering how businesses interact with customers by tailoring each touchpoint to personal preferences and anticipating needs before they even arise. What was once an impersonal, one-size-fits-all approach is being reimagined as AI helps to create highly customized encounters that make patrons feel like the singular focus.

The impact of this AI-driven renaissance in customer experience is monumental. By leveraging advanced data analytics to optimize the entire customer journey, businesses can forge deeper emotional connections, ignite brand loyalty, and elevate satisfaction to unprecedented heights. The future of exceptional, individualized customer service is already here—a new era of experience excellence unlocked by the incredible potential of AI to understand and cater to each person's unique needs and desires like never before possible.

THE AI-POWERED VISION FOR CUSTOMER EXPERIENCE

Have you ever wondered how Spotify's "Discover Weekly" playlist seems to know your music taste so intimately? That's AI at work, analyzing your listening habits, understanding your preferences, and curating a playlist that feels handpicked just for you. Or consider the online retail behemoth, Amazon. When you browse through its vast product listings, you're presented with recommendations that often align closely with what you're looking for, or even things you didn't know you wanted! Behind the scenes, AI algorithms are analyzing your browsing patterns, purchase history, and even the time you spend looking at products to predict and present what might catch your eye next.

How does this sound? You walk into your favorite coffee shop, and even before you place the order, the Barista who just joined the café already knows that you like almond milk and honey instead of sugar, and your top choice is Turkish aroma tea. She makes an instantaneous connection with you and is equipped to give you the best customer experience. AI is bringing this vision to life in customer experience. In today's digital age, customers are no longer satisfied with generic interactions. They crave experiences that resonate, that understand their preferences, and that anticipate their needs. With its unparalleled data processing capabilities and insights, AI is transforming the way businesses interact with their customers.

Remember the last time you had a query and turned to a website's chat support? Chances are, a chatbot, an AI-driven virtual assistant, assisted you by handling a myriad of customer queries in real time without human intervention. These chatbots are dramatically improving customer support by providing instant responses, reducing wait times, and ensuring customers get the answers they need when they need them.

Health-care chatbots guide patients on care options. Predictive analytics identify at-risk patients proactively. Intelligent virtual assistants in banking handle customer inquiries twenty-four-seven. AI predicts and meets evolving needs. Smart concierge like Marriott's AI-powered "RENAI By Renaissance" provides personalized recommendations.[1] Voice assistants enhance in-car experiences. Chatbots handle sales inquiries. AI analyzes usage patterns to recommend optimal cell phone plans. Churn prediction identifies at-risk customers. In Utilities, bots automated billing requests and outage communication. AI predicts demand. The applications are vast, but the goals are aligned—enhanced personalization, predictive engagement, and humanized customer journeys.

In essence, the vision for AI-powered customer experience is to create meaningful engagements that make customers feel valued, understood, and prioritized. Welcome to the future of customer experience, where every interaction is tailored, timely, and transformative.

CATALYST FOR CUSTOMER SEGMENTATION AND PERSONALIZATION

The advent of AI in customer segmentation and personalization has marked a transformative shift from traditional broad categorizations such as enterprise, channel, and direct segments, which often led to a generalized "one-size-fits-all" approach. In stark contrast, AI delves into the depths of customer data, enabling the creation of highly individualized segments and personas, moving well beyond the conventional demographic categories like "young adults" or "seniors." AI's prowess lies in its ability to identify and cater to nuanced segments such as "eco-conscious millennials" or "tech-savvy retirees," thereby allowing businesses to tailor their offerings with unprecedented precision.

Consider the scenario where AI sketches a detailed persona, say "Eco-friendly Emma," who is identified as a young adult with a passion for sustainability, frequently engaging with green products and advocating for environmental causes on social media. Armed with such granular insights, companies can craft marketing strategies, develop products, and curate experiences that resonate deeply with individuals like Emma, thereby fostering a sense of personal connection and loyalty.

AI-driven personalization goes beyond mere demographic targeting, advancing into the new world of hyper-personalization where every touchpoint with the customer is customized based on a rich tapestry of data including online behaviors, purchase patterns, and social media interactions. This level of personalization is exemplified by platforms like Netflix, which employs AI to offer highly personalized viewing recommendations, even tailoring the artwork displayed to each viewer based on their watch history.[2] Similarly, companies like Coca-Cola have harnessed AI for marketing campaigns and e-commerce experiences that dynamically adjust to the individual consumer's preferences and behaviors, significantly enhancing engagement and conversion rates.[3]

AI chatbots and virtual assistants make customer service more personal by having natural conversations that fit your situation. For instance, IBM's watsonx Assistant adapts to customer language patterns over time, making interactions more intuitive and effective.

However, the embrace of AI-driven personalization brings forth critical challenges, particularly concerning data privacy and ethical considerations. Businesses must navigate the delicate balance between delivering personalized experiences and respecting individual privacy. Transparency in data collection, informed consent, and responsible data usage are paramount to maintain customer trust and compliance with regulatory standards. Additionally, ensuring AI models are free from bias and offer personalization controls to customers allows for a more inclusive and respectful engagement, ultimately enhancing the value proposition of personalized services.

In conclusion, AI significantly enhances personalization and customer engagement, transforming business-customer relationships. By leveraging AI responsibly, businesses can achieve personalized experiences, increased customer satisfaction, loyalty, and growth, while maintaining privacy and ethical standards. This shift paves the way for the next frontier in AI-driven customer experience—virtual assistants.

VIRTUAL ASSISTANTS: YOUR DIGITAL CONCIERGES

Picture having a digital assistant with you twenty-four-seven, ready to answer your questions and handle your requests quickly. Chatbots and virtual assistants are like digital concierges, making your interactions with businesses much easier. Much like an attentive concierge at a high-end hotel, these AI-powered entities greet users with a warm virtual embrace, deftly handling tasks from scheduling appointments to offering tailored recommendations, ensuring no request goes unanswered and every interaction is a seamless, personalized experience.

The evolution of chatbots from mere script-following entities to AI-enhanced conversationalists marks a significant leap in customer service technology. These conversational AI agents can skillfully deduce customer requirements from scant details and craft responses precisely tailored to the individual. Virtual assistants have evolved into highly capable digital assistants. They can manage schedules, set reminders, and inject humor into interactions, all while becoming more intuitive and natural through advances in artificial intelligence.

Major companies like Google, Apple, and Spotify are harnessing the power of AI-driven chatbots and virtual assistants. Ever wondered how Spotify seems to know just the kind of song you're in the mood for? Or how Google Assistant can predict your next question? That's the brilliance of AI at work, analyzing user behavior and preferences to offer tailored solutions.

One of the most visible impacts of AI-driven personalization is in product recommendations. Online retailers use AI to personalize homepage layouts, ensuring that each visitor sees the most relevant banners, deals, and products. Even search results can be tailored, ensuring customers find what they're looking for faster and more efficiently.

Starbucks implemented Deep Brew, an AI-powered customer support platform to handle orders and questions. It can take orders by chat, voice, or gestures in their test stores. Deep Brew has enabled automated upselling, faster service, and a more personalized customer experience.[4] Sephora uses AI chatbots in their app to provide personalized skincare and makeup routines. Users can get product recommendations for their needs from the chatbot. Sephora also employs AI to match users' selfies to recommend specific makeup products and shades for their skin tone.[5]

The tangible benefits of AI-driven personalized virtual assistants are evident. Personalized campaigns yield higher engagement rates, and brands employing advanced AI strategies witness significant revenue boosts.

CUSTOMER ISSUE MANAGEMENT

The conference call droned on, but my attention was locked on the laptop's flickering screen. Without warning, it shut down. Frantically, I reconnected via phone, frustrated with my new laptop shutting down. After the call, I reached out to Asus, the laptop's manufacturer, hoping for a swift resolution. Little did I know it would be an odyssey of frustration. Shuffled from one agent to the next, I repeated my tale like a broken record—the serial number, purchase date, personal details. At least fifteen minutes of hold music looped endlessly between each transfer, fraying my nerves. Remote locations degraded the phone lines, garbling voices. Two agonizing hours later, I was no closer to a fix. The issue remained unresolved. Exhausted, I hung up, resigning myself to another battle another day. Many of us have encountered such a situation, where the simple act of seeking assistance becomes a test of patience.

Contrast this with the promise of AI-driven customer support, where an AI agent, empowered with natural language processing, intelligently discerns your issue and navigates through a trove of solutions using decision trees and rule-based systems. Companies like Apple are already harnessing machine learning to refine their AI assistants, making them more adept with each interaction.[6]

Recently I sat with John Michelsen, CTO of Krista.ai and discussed the transformative power of AI in customer service. He recounts the creation of Aria, an AI customer service agent for a health-care company, which expertly handled a patient's inquiry about his insurance coverage. Aria's proficiency stems from its ability to tap into extensive databases and CRM platforms, drawing from a wealth of data to provide not just answers but peace of mind.

The real magic of Aria lies in its use of generative AI to access and compile information from diverse sources, enhancing the speed and autonomy of customer service. This innovative approach not only streamlines the resolution process but also embodies a seamless blend of technology and human empathy, offering users the option to connect with a real person when needed.

AI systems use real-time analytics to track customer issues, processing large data volumes to segment customers based on current behaviors and predict future actions. These analytics are often powered by time series analysis, which allows the AI to predict future bottlenecks or delays in resolving customer issues, thereby preempting them.

Sentiment analysis, which AI can show in real time, allows companies to gauge customer emotions based on their interactions with the support system. This invaluable data can be used to further refine customer service strategies. As we look ahead, the capabilities of AI in issue resolution are set to expand exponentially. Predictive analytics and deep learning models could enable AI systems to identify and resolve issues even before they affect the customer. Think of your smart fridge diagnosing a cooling issue and ordering a replacement part, all before you even notice there's a problem!

The impact of integrating AI into customer service extends beyond enhanced satisfaction; it significantly lowers operational costs and provides rich insights into customer sentiments, paving the way for more refined and empathetic customer interactions in the future.

ADDITIONAL WAYS AI ENHANCES CUSTOMER EXPERIENCE

Technologies like stream analytics are being leveraged to analyze vast amounts of real-time data, offering personalized promotions instantly. Meanwhile, Generative Adversarial Networks (GANs) are creating immersive virtual shopping experiences by generating hyperrealistic images of products, allowing customers to see how clothes fit and move on virtual models tailored to their body size and preferences. Augmented reality (AR), powered by AI, complements this by enabling customers to visualize products in their own living spaces, further personalizing the shopping experience.

AI is advancing toward understanding and responding to human emotions through emotion detection, sentiment analysis, affective computing, and the development of empathetic chatbots. These technologies allow for more meaningful and natural interactions, building trust and rapport with customers.

FINANCIAL BENEFITS AND BUSINESS SUCCESS

The integration of artificial intelligence (AI) into customer experience strategies has shown significant financial benefits for businesses across various industries. Gartner forecasts that by 2026, conversational AI for customer service will reduce contact center labor costs by eighty billion dollars. According to other surveys, businesses that have implemented AI in their customer service have witnessed a 27 percent reduction in average handle time, improving customer satisfaction, and 34 percent have seen an increase in revenue.[7]

Area	AI Capabilities	Benefits
Personatilization	Segmentation, recommendations, personalized content	Increased satisfaction and engagement
Virtual Assistants	Natural language processing, dialog management	Instant, 24/7 support and services
Issue Resolution	Predictive analytics, sentiment analysis	Proactive issue identification and tracking
Emotional Intelligence	Emotion detection, empathetic conversations	Deeper customer connections
Customer Understanding	Clustering, pattern recognition, predictive analytics	Tailored experiences based on needs

Table 8: How AI Helps Customer Experience

CASE STUDY: TRANSFORMING CUSTOMER SERVICE AT OTP BANK WITH AI

OTP Bank, the leading banking group in Central and Eastern Europe, developed their own AI model—a Hungarian large language model (LLM) to enhance customer service and other functions. The LLM powers over multiple use cases, with an initial focus on customer interactions, fraud prevention, and security. One of the most exciting applications is improving customer service through conversational AI. OTP Bank built an intelligent virtual assistant that answers customer queries, provides support, and handles transactions via natural dialogs. This automates routine issues so agents can tackle complex problems.

For instance, the virtual assistant fields common questions about accounts, products, services, and more. By accessing customer data, it can respond to natural language queries such as current account balance, when next mortgage payment is due, and how to open a new joint bank account. The assistant also processes transactions like money transfers, bill pay, and loan applications autonomously. This simplifies how customers manage finances. Additionally, the LLM generates personalized recommendations based on customer histories, proactively resolves issues, and provides real-time support across channels. By automating repetitive tasks, the LLM boosts efficiency dramatically.[8] This allows human agents to focus on relationship-building and complex problem-solving.

Early results are promising. Customer satisfaction has increased thanks to quick, accurate responses and transactions. OTP Bank plans to expand the LLM's capabilities, ultimately transforming customer service.[9]

GETTING STARTED WITH AI FOR CUSTOMER EXPERIENCE

Implementing AI to transform customer experience requires careful planning and execution. Here are the best practices companies can follow. This builds on the framework we discussed in chapter 5.

ASSESSMENT AND PLANNING

- Conduct a thorough evaluation of your current customer experience operations, including analyzing various data points such as query types, volume patterns, customer satisfaction metrics, key performance indicators (KPIs), operational metrics, and people metrics.
- Clearly define your business objectives for implementing AI in customer experience, such as reducing costs, improving customer satisfaction scores, or optimizing operational metrics.
- Gain a comprehensive understanding of your existing technology landscape and tools, including call routing systems, self-service portals, chatbots, knowledge bases, and current automation solutions.

CUSTOMER JOURNEY MAPPING

- Identify the critical touchpoints throughout the customer journey where AI can potentially add value and enhance the overall experience.
- Determine the frequency and volume of interactions at these touchpoints to prioritize areas for automation.
- Plan for AI-led personalization strategies that can tailor the customer experience based on individual preferences, behavior, and context.

- Establish clear escalation paths for seamless human intervention when necessary, ensuring a smooth transition between AI and human agents.

TECHNOLOGY SELECTION

- Evaluate and select AI solutions that align with your specific needs and requirements while minimizing friction in the customer experience (CX) and employee experience (EX).
- Consider factors such as scalability, ease of use, total cost of ownership (TCO), and the ability to customize the AI solutions to your unique business needs.
- Ensure that the chosen AI solutions can seamlessly integrate with your existing systems, such as customer relationship management (CRM) platforms, knowledge bases, chatbots, and databases.
- Assess your infrastructure readiness and make necessary upgrades or modifications to support the successful deployment and operation of the AI solutions.

PILOT PROGRAM

- Start with a pilot that has a limited scope, allowing you to test and validate the AI solutions in a controlled environment. The pilot should have enough touchpoints to give a good set of lessons learned.
- Define clear metrics and KPIs to evaluate the AI system's performance during the pilot phase.
- Gather feedback and insights from the pilot program to identify areas for improvement and refinement.

TRAINING AND CHANGE MANAGEMENT

- Develop comprehensive training programs to upskill your staff and equip them with the necessary knowledge and skills to effectively utilize the new AI tools.
- Implement a robust change management strategy to facilitate smooth adoption and acceptance of the AI solutions within your organization.
- Address potential concerns, provide clear communication, and offer ongoing support to ensure a seamless transition for your employees.

SCALING AND OPTIMIZATION

- Leverage the insights and feedback gathered from the pilot program to refine and optimize the AI systems, addressing any identified issues or areas for improvement.
- Gradually expand the scope of AI self-service capabilities, extending the implementation to additional touchpoints and customer segments.
- Continuously monitor and analyze the performance of the AI solutions, making adjustments and enhancements as needed to ensure optimal customer experience and operational efficiency.

AI in CX Framework

Assessment & Planning
- Define business objectives
- Conduct a thorough evaluation
- Evaluate technology landscape

Customer Journey Mapping
- Identify critical touchpoints
- Determine interaction frequency and volume
- Plan AI-led personalization strategies
- Establish clear escalation paths

Technology Selection & Infra Readiness
- Evaluate and select AI solutions
- Ensure seamless integration with existing systems
- Access infrastructure readiness for AI

Pilots
- Start with a limited-scope pilot
- Define clear metrics and KPIs
- Gather feedback for improvement

Change Management
- Develop training programs for staff upskilling
- Implement change management strategy
- Provide support and clear communication

Scaling & Optimization
- Refine AI systems using pilot feedback
- Expand AI self-service capabilities
- Monitor and adjust AI solutions

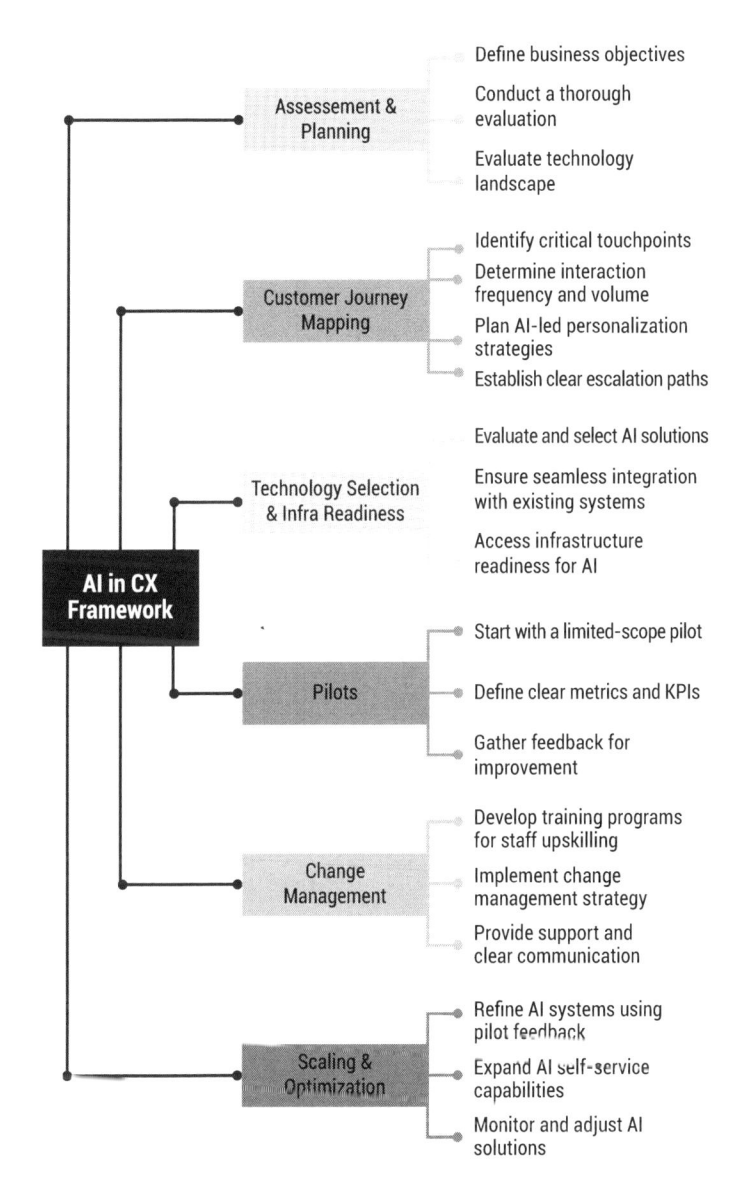

Fig 10: Steps to Bring AI to CX

AI is transforming customer experiences with personalized and intelligent interactions. It enables twenty-four-seven support through virtual assistants and proactive issue resolution with real-time analytics. Balancing personalization with privacy is crucial. Despite challenges, AI enhances customer engagement and loyalty. Thoughtfully applied, it enhances rather than replaces the human touch.

CHAPTER 12

The Art of AI: Redefining Sales And Marketing

What if you are craving a sweet treat in the morning? You open your fridge and see a tub of ice cream with a tempting label: "Ice Cream for Breakfast." You think, why not? It sounds fun and delicious. You scoop some ice cream into a bowl and enjoy it with a smile.

You may not realize it, but you have just been influenced by an AI-powered marketing platform called Persado. Unilever uses the platform to optimize its marketing efforts. The platform uses natural language processing (NLP) and machine learning to analyze over 1.5 billion interactions from 150 million customers and generate personalized marketing messages that appeal to different customer segments.[1] It finds what words, phrases, emotions, and images resonate with customers and persuade them to act.

For example, Persado discovered many songs mention "ice cream for breakfast" in the public domain, and this phrase evokes a sense of joy, curiosity, and indulgence. Based on this finding, Persado generated catchy slogans and headlines for

Unilever's new line of ice cream products, such as: "Ice Cream for Breakfast: The Ultimate Morning Treat," "Start Your Day with a Scoop of Happiness," "Break the Rules with Ice Cream for Breakfast." These messages were tested and optimized by Persado to ensure they deliver the best results.[2]

If you're a sales or marketing leader, you know the critical role sales and marketing play in the success of your organization. But have you considered the transformative impact of artificial intelligence in this domain?

REVENUE INTELLIGENCE: A NEW FRONTIER IN SALES

Revenue intelligence employs strategies, tools, and processes that leverage AI to analyze extensive data on customer behavior, market trends, and team dynamics. This data-driven approach transcends traditional sales metrics, offering actionable insights for strategic revenue optimization decisions. Companies like Aviso AI are pioneering this field, demonstrating the future of sales relies not just on relationships and intuition but on deep data analysis for smarter business decisions.

Generative AI with its underlying LLMs enables insights from unstructured data like never before. AI algorithms can sift through massive amounts of customer data enabling sales teams to understand customer needs and preferences at an unprecedented level, allowing for more targeted and effective sales strategies. One of the most significant advantages of AI in revenue intelligence is real-time decision-making. AI-powered platforms can provide real-time updates on customer interactions and market trends along with insights into customer behavior and sales performance, enabling sales teams to make informed decisions quickly. Forecasting sales has always been a bit of a guessing game, but AI is changing that. With advanced predictive analytics, AI can provide incredibly accurate sales forecasts, helping businesses to allocate resources more efficiently and strategize more effectively.

Aviso AI, a leader in revenue intelligence, has a unique and comprehensive approach to revenue intelligence. Their GenAI platform serves as an assistive front-end that helps revenue and go-to-market (GTM) teams deliver personalized decision-making at scale, offering a 360-degree view of customer information enabled by real-time CRM updates. Aviso employs its time series AI engine to integrate historical data with current deal progress, creating a unified view of your sales pipeline. This allows for more accurate and reliable sales forecasting.

AI can analyze sales calls, meetings, and even written communications to provide real-time feedback and coaching. This enables sales teams to understand what's working and what's not, allowing for continuous improvement and, ultimately, better sales outcomes. Aviso employs its conversational intelligence solution, known as MIKI, to improve a company's total annual revenue. It also helps to understand your team's performance to identify strengths and weaknesses, providing actionable insights for improvement.

Similarly, customer health scoring is currently a largely manual and subjective process, where customer success managers rely on subjective judgments combined with some rules-based approaches. AI is set to overhaul this by automatically creating health scorecards based on historical data, thus minimizing subjectivity. With a human-in-the-loop approach, customer success managers can still adjust these scorecards as needed, ensuring a balance between automation and expert oversight.

ENHANCING SALES WITH AI: FROM COLD CALLING TO CONVERSION

AI is transforming sales processes from initial prospect outreach to closing deals. For cold calling, AI lead scoring identifies the hottest leads to prioritize while personalized AI-driven emails and scripts increase engagement. AI also determines the optimal times for outreach based on data analysis. During warm calls, AI provides reps with comprehensive prospect insights for more meaningful conversations. Real-time AI coaching suggests tactics to better engage prospects, and automated follow-ups are scheduled based on interaction data.

When converting prospects to customers, AI analyzes multichannel behavior to pinpoint buying signals. Machine learning algorithms segment customers for hyper-personalized nurture campaigns tailored to their preferences. Follow-ups are automated based on prospect actions, guiding them smoothly through the funnel. AI-powered forecasting leverages historical and market data to forecast sales accurately, allowing timely, targeted engagement of potential buyers.

From initially identifying ideal prospects to ultimately closing deals, AI enhances sales effectiveness at every stage. By harnessing machine learning, contextual intelligence, behavioral data, and automation, AI allows sales teams to operate with unprecedented efficiency, personalization, and insight into customer needs and motivations.

CONTENT CREATION, PERSONALIZATION, AND OUTREACH

While earlier chapters explored how AI powers personalized content curation and recommendation engines, its capabilities extend far beyond that. AI models can create original, high-quality content tailored to specific audiences and objectives.

Rather than relying solely on human writers, companies can leverage AI assistants to rapidly generate blog posts, social media updates, video scripts and more. By analyzing data on content performance and trends, these AI tools can help predict what topics will resonate best and allow businesses to stay ahead of the curve.

Adobe's Sensei platform employs machine learning algorithms to analyze design trends and user preferences, offering real-time suggestions for improving content. It automates repetitive design tasks, freeing up creative teams to focus on more complex projects.[3]

AI is also transforming marketing outreach and lead conversion strategies. Gone are the days of batch-and-blast emails and cold calls. Machine learning can analyze a prospect's digital footprint—their online behavior, social media presence, and email communication style—to determine the optimal timing, channel and messaging for outreach. This personalized, AI-driven approach leads to interactions that feel natural and contextually relevant, rather than interruptive.

Beyond optimizing outreach, AI opens new frontiers in delivering truly individualized marketing across channels. Companies can use machine learning to dynamically tailor website content, email campaigns, ad creativity, and more to each customer's unique preferences and behaviors. This moves marketing from segmentation to individualization at scale.

SOCIAL MEDIA AUTOMATION AND ANALYSIS

As someone who frequently publishes articles and posts on LinkedIn, I once found the process of crafting long-form and short-form content an exhaustive endeavor. Ideating and writing, structuring paragraphs, selecting images, and finessing headings and subheadings consumed significant time, not to mention the occasional writer's block that stalled progress. However, the advent of generative AI tools like Claude, custom GPT agents, and Gemini has transformed this process for me. These AI assistants generate ideas, topics, and content outlines. They rapidly produce high-quality drafts tailored to audience preferences and even assist with editing, restructuring, and formatting. As a result, my productivity was up by 50–70 percent, allowing me to dedicate more time and energy to strategic tasks like promotion, networking, and building thought leadership on LinkedIn.

Social media has emerged as a pivotal arena for brand development, customer interaction, and lead conversion. The challenge, however, lies in the intricate management of various platforms, each demanding constant content updates, engagement monitoring, and performance analysis. This is where the prowess of AI becomes indispensable, automating routine tasks such as scheduling posts across multiple channels, replying to commonly asked questions or mentions, and uncovering insights that can redefine your digital strategy.

One of the critical issues in social media management is handling the vast expanse of user-generated content, which sometimes veers into the territory of being abusive or illegal. AI steps in as a vigilant overseer, automating the detection and flagging of such content, thereby making moderation both efficient and scalable.

Furthermore, AI's capabilities extend to content scheduling and optimization. Tools equipped with machine learning, such as Hootsuite, delve into historical engagement data to recommend optimal posting schedules. This not only ensures your content achieves greater visibility but also fosters enhanced interaction with your audience, aligning with your brand's voice and audience's preferences seamlessly.

Understanding customer sentiment is crucial for brand management. AI can analyze comments, likes, shares, and emojis to gauge public opinion about your brand. This enables you to tailor your social media strategy to resonate with your audience. Sprinklr uses natural language processing to analyze customer comments and reviews, providing real-time insights into customer sentiment and trends.[4]

Finding the right influencers to collaborate with can be a hit-or-miss endeavor. AI algorithms can analyze influencer metrics such as engagement rate, follower demographics, and content quality to identify the best influencer for your brand. Upfluence employs AI to scan through a database of influencers, matching brands with influencers who align with their target audience and brand values.[5]

Measuring the ROI of social media campaigns can be complex, but AI simplifies this by providing advanced analytics that go beyond likes and shares. Think of customer lifetime value, conversion rates, and even predictive analytics for future campaigns. Socialbakers uses machine learning to provide comprehensive analytics—from engagement metrics to ROI calculations—helping brands make data-driven decisions.[6]

AI moderation offers a powerful solution for social platforms to effectively manage the immense volume of daily user posts, a task that would be overwhelming for human reviewers. By leveraging AI, platforms can rapidly and efficiently filter out objectionable content, enhancing user experience. Additionally, AI moderation alleviates the emotional toll on human moderators, who would otherwise be repeatedly exposed to disturbing or traumatic material over extended periods.

However, it is crucial to acknowledge that AI moderation is not infallible and requires constant training, iteration, and human oversight to improve accuracy. Platforms must implement robust appeals systems to address instances where users believe their content has been incorrectly flagged. Ultimately, AI moderation empowers social platforms to combat abuse and misinformation at scale while reducing the burden on human moderators, striking a balance between content moderation and user experience.

Facebook uses AI tools to detect nudity, hate speech, terrorist propaganda, and other harmful content in posts, images, videos, live streams, and comments. Algorithms flag suspicious content for human reviewers.[7] YouTube uses deep learning AI to identify violent extremist content and automatically remove videos that violate its policies. Machine learning helps detect inappropriate comments.[8] TikTok's AI moderation reviews videos, captions, and live streams to identify adult nudity, illegal activities, and violent or graphic content.[9] This helps maintain TikTok's focus as an app for shared joyful experiences.

AI has become a powerful tool for content creation on social media. In the next section, we will explore how AI can be used for SEO optimization and digital campaigns, another crucial aspect of a successful online presence.

SEO OPTIMIZATION AND DIGITAL CAMPAIGNS

Effective SEO optimization and dynamic digital campaigns are pivotal components of a successful marketing strategy. As the complexity of the online world grows, AI is emerging as a game-changing solution, offering innovative and incredibly effective approaches to these critical areas.

Modern SEO revolves around understanding user intent and delivering content that not only ranks highly but also resonates with the target audience. AI goes beyond just analyzing on-page ranking factors. It can make smart suggestions to optimize page elements. By analyzing thousands of pages, AI can automatically optimize elements like header tags (H1, H2, etc.) and image filenames for better keyword targeting and on-page SEO, boosting search visibility and click-through rates.

Voice search introduces new SEO complexities with its conversational nature. AI can identify opportunities to optimize pages for long-tail voice queries and generate content tailored specifically for voice search results, allowing brands to tap into the growth in voice search traffic.

AI also enables real-time tracking of campaign performance with actionable insights. Instead of waiting for reports, marketers can leverage AI to monitor conversions, engagement, costs, and other KPIs in real time while also detecting shifts and recommending optimizations to improve results, enabling continuously optimized campaigns.

AI solutions like Moz, Semrush, and TapClicks are being leveraged by top brands for SEO and campaign optimization, providing a robust framework to operationalize AI across the digital marketing stack.[10] Marketers can leverage capabilities like predictive lead scoring, social media optimization, and campaign management powered by AI to boost ROI.

AI-driven platforms can provide in-depth keyword analytics, helping businesses understand their digital ecosystem positioning. Advanced AI algorithms can analyze on-page ranking factors in real time, enabling marketers to optimize content as they go, and automate repetitive SEO tasks like backlinking and content optimization, making the process more efficient and effective.

The integration of AI into SEO optimization and digital campaigns is setting a new standard for what's possible in digital marketing. It's not just about collecting data. It's about making intelligent decisions based on that data, driving better results and a more personalized, engaging experience for audiences.

	Basic AI	
Automated Content Ideas Content Optimization Search Trend Analysis		Campaign Performance Tracking Social Media Optimization Demographic Targeting
SEO Optimization		**Digital Campaigns**
Keyword Analytics On-Page Analysis Voice Search Optimization	Advanced AI	Forecasting Content Engagement Real-time Personalization Consumer Behavior Analysis

Fig 11: How AI Can Help SEO Optimization and Digital Campaigns

CASE STUDY: UNLOCKING THE POWER OF CUSTOMER INSIGHTS

Burnley Football Club, steeped in over 140 years of tradition, sought to transform its Northwest England roots into an internationally recognized brand. To engage fans worldwide and drive revenue beyond match tickets, the club required sophisticated social media capabilities. However, manually monitoring brand sentiment, benchmarking against rivals, providing unified analytics across channels, and tailoring content proved daunting within their budget constraints.

The club leveraged Sprinklr's Social Advanced platform, powered by cutting-edge AI technology. AI-driven caption tailoring ensured posts resonated with the voice and tone of each social platform. AI-enabled competitive benchmarking compared Burnley's performance to other clubs. Social listening AI analyzed online conversations to surface nuanced insights into brand perception. Additionally, AI analytics delivered real time, unified reporting on post-performance and audience demographics across channels to justify ROI to potential sponsors.

Leveraging Sprinklr's AI capabilities proved transformative for Burnley FC's global brand amplification efforts. Within just thirty days, AI-optimized posts drove a 264 percent increase across social channels along with 54 percent higher engagements. Media coverage grew 53 percent with 133 percent more top-tier publication mentions. AI-powered insights allowed refinement of Burnley's content strategy for deeper fan connections worldwide.[11]

AI in Sales & Marketing

- **Revenue Intelligence**
  - 360-degree Customer Insights
  - Predictive Analytics for Sales Forecating
  - Real-time Coaching for Sales Teams

- **Content Creation**
  - Automactic Content Creation
  - Data Analysis for Audience Preferences

- **Personalization**
  - Customer Analysis
  - Personalized Marketing
  - Product Recommendations

- **Sales Outreach**
  - Optimal Timing
  - Channel Selection for Sales
  - Prospect Behavior Analysis

- **SEO**
  - On-Page Optimization
  - Keyword Analytics

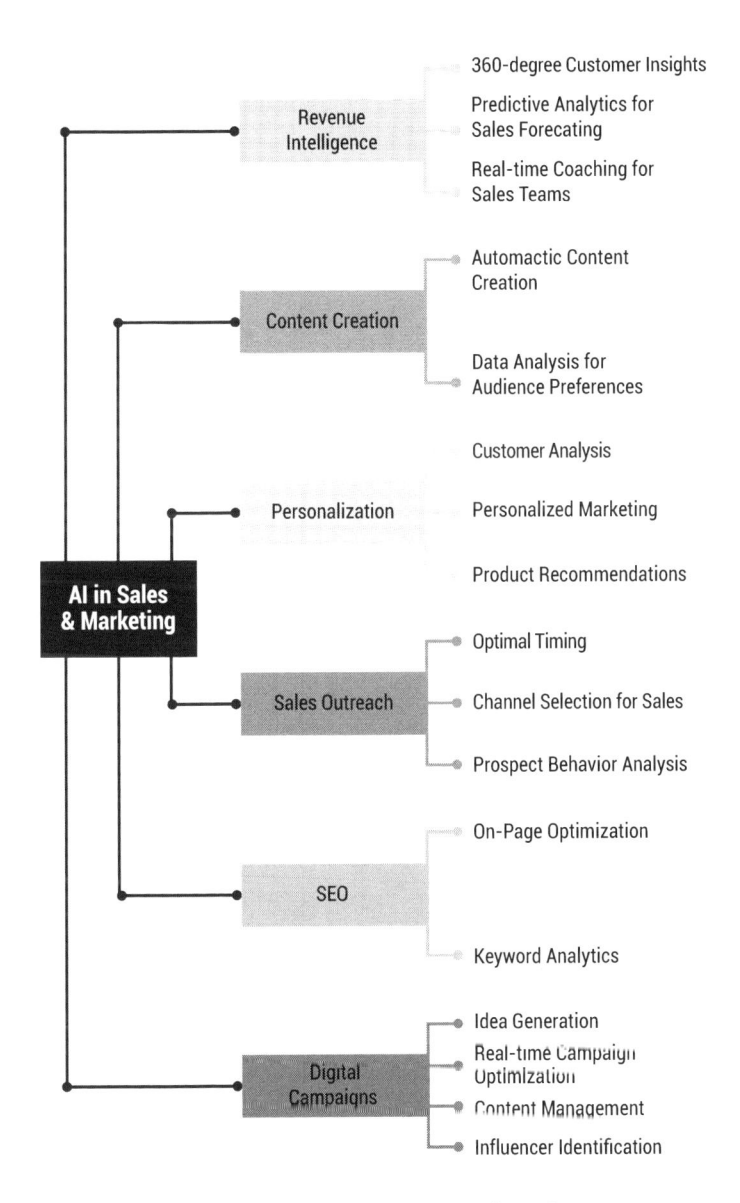

- **Digital Campaigns**
  - Idea Generation
  - Real-time Campaign Optimization
  - Content Management
  - Influencer Identification

Fig 12: How AI Can Help Sales and Marketing

AI is radically transforming sales and marketing, driving efficiencies and personalized experiences across the entire funnel. From revenue intelligence and forecasting to content creation, social media automation, and SEO optimization, AI infuses data-driven insights and intelligent automation. Early adopters leveraging AI for hyper personalized engagement are realizing competitive advantages. As the technology evolves, embracing AI will be critical for sales and marketing success in our customer-centric era.

CHAPTER 13

Operations Overhaul: AI-Driven Transformation

Nucor Steel makes a million tons of steel a year. With a diverse product range—from small components like paper clips to large-scale items like crane cables—the company struggled to maintain consistent quality while maximizing production efficiency. The training of new operators was time-consuming and resource-intensive, often taking months to years, leading to bottlenecks and inconsistencies in production quality. This company, with its rich history in steel production, has always been at the forefront of adopting new technologies. However, the complexity and variability in their production processes posed a unique challenge. The company's commitment to safety and quality, coupled with its ambition to lead in the steel industry, set the stage for a transformative journey.

The solution came in the form of Delta Bravo AI, a sophisticated AI system designed to enhance both workforce readiness and product quality. Delta Bravo AI was used to simplify operator decision-making, reduce training time, and drive higher quality outcomes. The AI system analyzes live data and predicts optimal solutions for a myriad of grade combinations. This has led to a paradigm shift in how Nucor approached its operations, significantly reducing the training period for new operators by enabling them to make decisions with the precision and understanding of seasoned veterans. The implementation of AI led to a remarkable transformation within the company. Training times were drastically reduced, operational efficiency skyrocketed, and most importantly, the quality of the products reached new heights of consistency and reliability.[1]

If you're an operations leader, you're no stranger to the complexities of running a smooth operation. Wouldn't it be wonderful if your inventory management is so accurate that stockouts and overstocks are a thing of the past? Or your order fulfillment process is so efficient that customers get their packages even before they expect them? Well, with AI, you could realize this potential.

WORK PRIORITIZATION AND MANAGEMENT

The ability to effectively prioritize tasks based on their urgency and importance is crucial for efficient work management. In manufacturing settings, AI algorithms are redefining task prioritization, particularly in maintenance operations. By analyzing equipment data and production schedules, AI can anticipate potential failures and prioritize maintenance activities for critical machinery during off-peak hours, minimizing disruptions to production cycles. The construction industry benefits from AI's capability to prioritize tasks based on resource availability, project deadlines, and milestone dependencies, enhancing overall project efficiency and adherence to schedules.

In agriculture, AI plays a pivotal role in scheduling planting and harvesting activities, guided by data on weather patterns and crop conditions, enabling more informed decision-making and maximizing yields. Health-care organizations are employing AI to transform nurse scheduling processes, considering a multitude of factors such as staff skills, patient needs, and workload balancing. This leads to more efficient operations and improved patient care outcomes.

In the consulting industry, AI is facilitating the matching of employee skills and expertise with project requirements, optimizing staffing processes and aligning personnel with their career aspirations. The fast-paced e-commerce sector is leveraging AI-driven dashboards that provide real-time insights into inventory levels and customer behaviors, enabling swift and informed decision-making to meet fluctuating demands.

Underpinning these AI implementations are advanced machine learning algorithms that excel at discerning patterns and trends from historical data, enabling accurate predictions for future scheduling and resource allocation. Neural networks in project management can anticipate potential delays and suggest mitigation strategies, continually improving their performance with exposure to more data. In the logistics industry, reinforcement learning techniques are employed to optimize delivery routes over time, enhancing efficiency and reducing operational costs. Natural language processing (NLP) plays a vital role in extracting critical information from project-related documents, identifying key milestones, dependencies, and requirements, and facilitating more informed decision-making.

AI shows a big potential for work optimization and operational efficiencies. A case study by Accenture reveals that leveraging AI automation led to a 35 percent cost savings.[2] IBM's adaptive workload management, powered by AI, has resulted in 30 percent database performance improvements.[3] These figures underscore the potential of AI to streamline operations, boost organizational efficiency, and drive tangible business value.

FORECASTING AND CAPACITY PLANNING WITH AI

Accurately predicting future demand has long been a challenge for businesses across industries. Fluctuations in customer demand can lead to costly excess inventory, staffing issues, or disappointing stockouts and lost sales. However, recent advances in AI and machine learning are transforming demand forecasting capabilities.

As the former COO of a recruitment outsourcing business, I experienced this forecasting challenge firsthand. With over four hundred thousand annual requisitions, we lacked visibility into demand patterns. Our reactive approach of scrambling to hire temps whenever a new client came onboard proved inefficient and costly. By the time we onboarded and trained new workers, demand had often shifted again.

The solution was implementing data-science-driven forecasting models that could analyze historical requisition data, economic indicators, job market trends, and other relevant factors to predict staffing needs accurately. This allowed us to proactively plan talent acquisition instead of reacting to short-term fluctuations. The AI system continuously learned and adapted its forecasts as new data emerged.

The benefits were substantial—smoother staffing ramps, reduced overstaffing costs, and increased margins by eliminating inefficiencies from reactive hiring. We could finally align our workforce supply intelligently with data-driven demand forecasts.

AI forecasting distinguishes itself through intelligent model ensemble techniques and adaptability. By combining predictions from multiple models tailored to the data's size and complexity, AI can optimize accuracy. Simple models handle smaller datasets while complex deep learning models discern nuanced patterns in large data volumes.

Moreover, AI can ingest and learn from data streams in real time, a major advantage when demand patterns are volatile. This adaptive capability is invaluable across industries like manufacturing, retail, hospitality, energy, and utilities where demand is influenced by diverse fluctuating variables.

According to McKinsey research, applying AI forecasting reduces errors to supply chains by 20–50 percent and call center volume forecasting misses by 10 percent. This translates into up to 65 percent less lost sales and product stockouts, 5–10 percent lower warehousing costs, and 25–40 percent reduced administrative expenses. Entire workforce management processes can be automated, driving 10–15 percent cost savings while improving operational resilience.[4]

From forecasting energy needs based on weather and occupancy rates to predicting beverage sales down to the SKU level using economic and social signals—AI synthesizes disparate data sources to generate incredibly accurate demand forecasts. This precision empowers smarter inventory management, staffing optimization, and strategic decision-making across business operations.

AI IN SUPPLY CHAIN AND LOGISTICS

As we have seen during COVID-19 and beyond, supply chain and logistics are more than just the backbone of any business; they're the competitive edge and have the potential to alter economic landscape across countries. Managing a supply chain has always been complex, juggling variables such as inventory levels, supplier reliability, and transportation logistics. But what if AI could manage this complexity and turn it into a strategic advantage?

A variety of AI techniques are transforming supply chain operations. Machine learning algorithms detect patterns in inventory data to accurately forecast demand. As they process more data, these algorithms continuously fine-tune their models to improve predictions. Deep learning neural networks identify anomalies in supplier networks, enabling mitigation of risks. Reinforcement learning optimizes routing and logistics by learning through trial-and-error interactions with the environment. As these AI models train on new data, their decision-making becomes more robust over time.

Rakuten Super Logistics, a leading e-commerce fulfillment company, implemented AI-powered autonomous mobile robots (AMRs) from Zebra Technologies to boost productivity and efficiency in their warehouse operations. By integrating Zebra's FlexShelf AMR system with their warehouse management software, Rakuten was able to improve picking efficiency by reducing the time pickers spent walking by up to 60 percent, increase order fulfillment accuracy through pick-to-light systems that simplified training, and enhance worker safety—all while delivering up to three times higher productivity compared to manual fulfillment processes.[5]

Keeping track of inventory across a retail chain with multiple locations can be a logistical nightmare. But with AI algorithms, not only can you monitor stock levels in real time, but you can also track your shipments continuously using GPS and IoT device data. AI can sift through data points ranging from geopolitical factors to historical performance to assess supplier risk. But that's not all. AI can also analyze past sales data, market trends, and even social media chatter to forecast customer demand with astonishing accuracy. This enables you to align your supply chain operations seamlessly, reducing both overstock and stockouts.

Fuel costs can make or break your shipping business. AI-powered systems can analyze traffic patterns, weather conditions, and fuel prices to suggest the most cost-effective routes. This not only saves time but also significantly reduces fuel costs and carbon emissions.

Here are a few companies leading the AI advancements in supply chain and logistics. Echo Global Logistics utilizes AI and machine learning technologies in several key ways to optimize supply chain management for its customers. It leverages data science and advanced algorithms to provide accurate pricing predictions, which shippers and carriers can access across Echo's platforms to aid in budgeting and forecasting. AI and machine learning also help automate mundane tasks, allowing Echo's human employees to focus on more complex issues and building strong relationships with partners. By intelligently combining AI capabilities with human expertise, Echo is able to effectively manage over sixteen thousand shipments daily while delivering tailored technology solutions across its large network.[6]

Amazon utilizes several AI technologies, including deep learning, natural language processing, and computer vision, to optimize packaging decisions and reduce waste in its supply chain operations. A machine learning model analyzes text data like product names and descriptions alongside visual data from computer vision tunnels to determine the optimal packaging type and dimensions for each product. This AI-driven approach has enabled Amazon to reduce per-shipment packaging weight by 36 percent over six years while eliminating over a million tons of packaging waste. By leveraging AI to make data-driven packaging choices at scale, Amazon can minimize unnecessary materials, lower costs, and increase customer satisfaction.[7]

THE POWER OF AI IN DEMAND FORECASTING

Accurately predicting future demand has long been a major challenge for businesses across industries. Fluctuations in customer demand can lead to costly excess inventory or disappointing stockouts. However, recent advances in AI are transforming demand forecasting capabilities. By harnessing machine learning and deep learning models, companies can now generate far more accurate demand predictions to drive smarter operational and strategic decisions.

Fast food giant McDonald's provides a powerful example of leveraging AI to crack the demand forecasting code. When Steve Easterbrook took over as CEO in 2015, he recognized the company was serving billions of burgers annually with little insight into their diverse customer base. Easterbrook kicked off an ambitious digital transformation centered on better understanding customers through data.

This required modernizing McDonald's outdated on-premises data infrastructure. Recently, I caught up with Gokula Mishra, the leader of data and analytics at McDonald's at that time. Under Mishra's leadership, the company completed a rapid full migration to the cloud in eight months. The new centralized, scalable cloud data platform enabled McDonald's to finally gain a 360-degree view of customers, supply chain, and operations.

With the data foundation in place, AI innovation took hold. Computer vision reduced drive-thru order errors by up to 80 percent in many restaurants. Supply chain algorithms predicted demand with far greater accuracy to optimize inventory and reduce waste. Dynamic scheduling improved labor efficiency and franchisee profits. Marketing AI personalized meal deals to individual customers based on purchasing patterns.

Through this transformation, McDonald's evolved from having virtually no customer insights to embedding AI-driven demand forecasting and optimization across its global business. As Mishra stated, "Now we can see everything."

AI-driven demand forecasting allows businesses like retailers and restaurants to optimize inventory distribution, staffing, and marketing tactics by accurately anticipating demand fluctuations. AI demand forecasting also guides strategic decisions by surfacing emerging consumer preferences before competitors.

In the dynamic e-commerce space, AI demand forecasting allows companies to implement more intelligent pricing strategies. By continually ingesting data on market dynamics, an AI system can automatically adjust prices in real time based on anticipated demand shifts for different products. This maximizes revenue while passing savings to customers through timely promotions.

Beyond optimizing day-to-day operations, AI-powered demand forecasting provides a strategic advantage by surfacing emerging consumer trends and preference changes before competitors. Businesses can stay ahead of the curve by proactively aligning product roadmaps, inventory plans, and marketing strategies with anticipated demand patterns identified by AI.

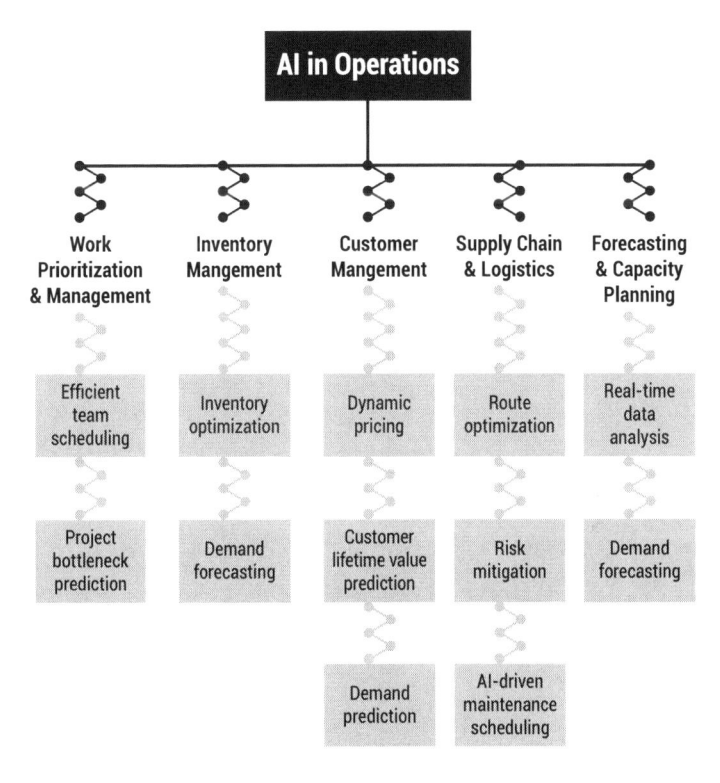

```
                        AI in Operations
    ┌──────────┬──────────┬──────────┬──────────┐
   Work     Inventory  Customer  Supply Chain  Forecasting
Prioritization Mangement Mangement & Logistics  & Capacity
& Management                                    Planning

  Efficient   Inventory   Dynamic     Route      Real-time
    team    optimization  pricing  optimization    data
  scheduling                                     analysis

   Project     Demand    Customer     Risk        Demand
  bottleneck forecasting lifetime   mitigation  forecasting
  prediction            value
                        prediction

                         Demand     AI-driven
                        prediction  maintenance
                                    scheduling
```

Fig 13: Operational Areas Ripe for AI

As AI capabilities continue advancing, their impact on demand forecasting will intensify. Companies able to effectively harness and scale this transformative technology will gain an unmatched ability to anticipate and meet customer demand proactively across their operations.

CHAPTER 14
AI Transformation in Human Resources: From Recruitment to Retention

---

Hilton Hotels embarked on a transformative journey in recruitment, leveraging the AI system AllyO. Hilton's challenge was a familiar scenario in the hospitality industry—the overwhelming task of processing numerous job applications. Traditional methods, often labor-intensive and lacking in diversity, pointed to a pressing need for a more streamlined and equitable recruitment strategy.

AllyO, an AI tool functioned as an interactive recruitment assistant, engaging applicants through text and web chats to mark a significant shift from conventional recruitment methods. Through these interactions, AllyO assessed candidates' suitability for various roles, bringing a new level of personalization to the recruitment process. Beyond initial interactions, AllyO's functionalities extended to interview scheduling, reminder dispatch, and feedback collection, effectively reducing administrative workload and enriching the recruitment experience.

The deployment of AllyO yielded impressive results for Hilton, providing first-class candidate experience, and automated administrative tasks. The impact is astonishing as they're now able to send 83 percent more offers per recruiter per week.[1]

In AI-enhanced recruitment, prioritizing the elimination of biases and strict adherence to safety and governance standards is essential. AI's efficiency and analytical power must be balanced with vigilant human oversight to prevent the perpetuation of existing biases. Integrating a human-in-the-loop approach ensures that AI's capabilities complement, rather than replace, ethical human judgment. Consistent audits and transparent operations are crucial in maintaining unbiased and accountable AI systems as we discussed in chapter 9 "AI for Good: Developing AI Responsibly and Ethically."

AI-POWERED PAYROLL

Managing payroll has always been a challenging task filled with complex, error-prone manual processes, tedious data entry, and the risk of noncompliance with changing regulations. The advent of AI is dramatically changing payroll management, ushering in a new era of efficiency, accuracy, and strategic insights by enabling AI-driven algorithms to automate intricate computations, such as salary deductions, tax calculations, and benefit deductions with a high degree of precision. By leveraging machine learning models trained on vast datasets encompassing payroll data, tax codes, and compliance rules, these AI systems can process complex calculations more accurately than traditional manual methods, significantly reducing the risk of errors.

AI also introduces intelligent mapping and data standardization. AI algorithms can automatically map general ledger codes and standardize payroll data from diverse sources, ensuring consistency and integrity in financial reporting. This feature is particularly valuable for multinational organizations dealing with various payroll systems and regulatory environments across different countries.

AI-powered dashboards offer unprecedented visibility into payroll operations, enabling in-depth analysis and the generation of actionable insights to support strategic decision-making and improve overall operational efficiency.

One of the most groundbreaking applications of AI in payroll is anomaly detection, which significantly bolsters fraud prevention efforts. Sophisticated AI algorithms can learn from millions of payroll runs to identify normal activity patterns, enabling them to detect anomalies or outliers that could indicate potential fraud. This proactive approach to security enhances the robustness of payroll systems, safeguarding sensitive financial data and mitigating risks.

Furthermore, AI systems can ingest and process complex regulations and business logic that would overwhelm traditional systems. This capability, combined with continual learning, ensures AI-powered payroll solutions remain current with the latest regulatory changes without relying on infrequent manual updates, thus reducing the risk of noncompliance.

Real-world case studies, such as those of Kirby Group Engineering and Cloudera, exemplify the tangible benefits of AI-powered payroll solutions.[2] From automating data processing and calculations to enabling intelligent mapping, enhancing analytics, and streamlining integration with HCM systems, AI is alleviating a major operational burden for HR departments across industries.

As organizations embrace the power of AI in payroll management, they are poised to experience significant time savings, reduced manual labor, improved operational efficiency, and enhanced strategic decision-making capabilities.

THE DAWN OF A NEW ERA IN TALENT ACQUISITION

A few years ago, before AI significantly impacted the landscape, I managed a recruitment process outsourcing (RPO) business at ADP. Each time we onboarded a new client, our team—spanning account services to operations—would need to scale up significantly, often requiring eight to ten full-time employees ranging from recruiters to back-office support. This approach was hardly sustainable and certainly not scalable.

AI has revolutionized recruitment by streamlining tasks and enhancing efficiency while essential human oversight ensures fairness and ethical standards, as seen with Hilton's use of AllyO.

AI is making a big impact on recruitment by creating sharper, more informative job descriptions and requisitions. Traditionally, this has been a manual, time-consuming task, requiring a significant amount of a recruiter's bandwidth. Now, with the advent of sophisticated AI tools, it is possible to generate detailed job posts swiftly and efficiently. These AI solutions analyze a myriad of data points—including historical requisitions, employee profiles, and performance evaluations—to produce job descriptions that accurately capture the necessary skills, responsibilities, and characteristics for specific roles.

The implementation of AI in this domain could significantly reduce the time spent on drafting requisitions, 50–80 percent based on my discussions with some recruiters who are using AI tools. This shift in workload allows recruiters to redirect their focus toward more strategic activities including enhancing employer branding, expanding professional networks, and building meaningful relationships with candidates and hiring managers.

Another key application of AI is in candidate screening. Powerful algorithms can scan thousands of résumés in seconds, identifying the most qualified applicants by analyzing skills, experience levels, and even cultural fit with the company. This automated screening represents a massive time-saver for recruiters and HR professionals.

In addition to screening, AI is being used for automated interview scheduling. AI-powered chatbots can communicate with candidates to find the best times for interviews that work for both parties. This eliminates the frustrating back-and-forth traditionally required to coordinate schedules. As a result, recruiters are freed up to focus on more meaningful and strategic hiring initiatives instead of calendaring logistics.

AI is also transforming recruitment through predictive analytics. By compiling and analyzing a wide range of data points—from a candidate's past performance to psychometric test results—AI can forecast future job success with impressive accuracy. This allows companies to better evaluate candidates and make smarter hiring decisions.

Leaders in the AI-enabled recruitment sphere, such as HireVue and Pymetrics are at the forefront of these innovations. HireVue's technology analyzes video interviews, evaluating facial movements, word choice, and voice to compare candidates' results with high achievers. It measures competencies such as communication, empathy, teamwork, and decision-making skills.[3] Pymetrics uses neuroscience-based games and AI to assess candidates' traits, matching them with top performers at companies. Their de-biased algorithms focus on skills rather than demographics, enhancing fairness, diversity, and recruitment efficiency while saving resources.[4]

LEVERAGING AI FOR RETENTION

As organizations strive to maintain a competitive edge in the fast-paced business landscape, retaining top talent has become a top priority. AI offers a powerful solution to this challenge by enabling strategic workforce planning and targeted retention efforts.

Through predictive analytics, AI can identify high-potential employees who are at risk of attrition based on factors such as performance metrics, engagement levels, and career progression patterns. This early warning system empowers HR leaders to proactively intervene with personalized retention strategies, such as tailored development opportunities, mentorship programs, or compensation adjustments.

Furthermore, AI can analyze employee data to uncover the key drivers of engagement and job satisfaction among top performers. By understanding the factors that contribute to their success and fulfillment, organizations can refine their HR practices to create an environment that nurtures and retains their most valuable assets.

AI-powered sentiment analysis can also provide invaluable insights into employee sentiment by analyzing internal communications, surveys, and feedback. This enables HR teams to identify potential pain points or areas of dissatisfaction before they escalate, allowing for timely interventions and continuous improvement of the employee experience.

THE AI-DRIVEN ONBOARDING AND TRAINING ASSISTANCE

This story was shared by Sarah (name changed) in a conversation. Sarah was a superstar at her previous company and was promoted quickly. She was excited to start her new job as a finance manager at a software company in San Francisco. Since everyone was remote during COVID, Sarah looked forward to connecting with her manager,

Michael, over video meetings during the onboarding process. However, Michael was extremely busy and rarely had time to talk to Sarah.

The company used a legacy financial system that few people knew how to use. Sarah was simply given a list of tasks with no training on the complex software she was expected to use. She felt overwhelmed trying to figure everything out on her own. When she asked questions, Michael told her to search the company wiki or put in support tickets.

Without a proper onboarding program or manager's support, Sarah quickly became frustrated. She dreaded logging in each day to battle the convoluted financial system alone. The isolation and stress began to negatively impact Sarah's mental health. After six agonizing months, she quit. She wondered if they even cared about retaining new hires like her.

The first few weeks at a new job can be overwhelming for employees. From understanding company policies to getting up to speed with job-specific skills, the onboarding and training process is critical. But what if this process could be made more efficient, personalized, and engaging? That's where AI can help.

We all have been there—as a new hire we have a million questions from IT setup to benefits and processes. Instead of waiting for a human response, AI-powered chatbots can provide instant answers from a company's knowledge base, making the onboarding process smoother and less stressful. These chatbots can be programmed to handle a wide range of queries—from simple administrative questions to more complex job-related inquiries.

AI doesn't just stop at onboarding. It extends into ongoing training as well. By analyzing performance metrics and feedback, AI can predict which employees might struggle with certain tasks or roles. This enables HR to proactively offer additional training or resources, ensuring employees are not just onboarded but are also continuously supported in their career growth.

Leading HR platforms are pioneering this AI-enabled experience by integrating advanced technologies to create a more dynamic and responsive onboarding process. As an example, Zavvy utilizes AI to tailor onboarding experiences to individual needs and learning styles.[5] Their platform analyzes employee data to personalize training materials and match new hires with mentors.

PERFORMANCE REVIEWS AND FEEDBACK ANALYSIS WITH AI

During my Intel days, annual performance reviews and feedback was the most dreaded time for both managers and employees. The process, known as Focal, was brutal and was derisively referred to as "ranting and raving."

Performance reviews, while fundamental to HR practices, have long been challenged by issues of subjectivity, bias, and the difficulty of providing constructive feedback. However, with the advent of AI, we're seeing a paradigm shift in how these reviews are conducted, offering a more positive experience with a growth-focused approach.

AI tools like Anthropic's Claude provide a framework for feedback that is developmental and attuned to individual growth needs. By analyzing past performance data, team dynamics, and individual feedback responses, these AI solutions craft nuanced, effective feedback, streamlining the process for managers and enhancing its reception among employees.

The introduction of AI into performance reviews brings a new level of objectivity and fairness, analyzing written or spoken feedback to minimize bias. This transparency not only refines the review process but also makes it more insightful and equitable. Additionally, AI's predictive capabilities extend beyond mere evaluation, playing a crucial role in career development. By examining various data points like job performance and skill sets, AI can forecast potential future roles and career paths for employees. This allows HR and managers to provide targeted development plans, focusing on nurturing employees for future success rather than merely evaluating their past performances.

AI's role in performance reviews is multifaceted, involving techniques such as natural language processing for sentiment analysis, machine learning for identifying trends, conversational AI for real-time feedback, and text generation for drafting personalized review drafts. This comprehensive approach not only saves time but also ensures the feedback is well-rounded and tailored to individual needs.

The evolution of performance assessments is vividly illustrated by initiatives from companies like BetterWorks and Plai. BetterWorks harnesses the power of generative AI to craft detailed and impartial evaluations of employees, drawing from their complete contributions throughout the review period.[6] Meanwhile, Plai introduces AI-driven tools that swiftly analyze reports, offer intelligent recommendations for objectives and feedback, and unveil key insights about your company and its teams, enhancing strategic decision-making and organizational understanding.[7]

Ultimately, the integration of AI in performance reviews aims to transform the process into a nurturing, empowering experience that motivates and develops employees. By leveraging AI's advanced analytical capabilities and personalized approach, performance reviews evolve into a catalyst for growth and success, aligning employee aspirations with organizational goals.

AI-ENABLED LEARNING MANAGEMENT SYSTEMS

Have you come across a corporate training program that knows exactly what you need to learn, when you need to learn it, and how you learn best? AI is transforming learning management systems (LMS) from mere repositories of training materials into dynamic, adaptive platforms.

Traditional LMS platforms often offer a one-size-fits-all approach to training, which can be ineffective and demotivating for employees. AI on the other hand analyzes individual learning styles, performance metrics, and career goals to create personalized learning journeys.

AI-enabled LMS platforms can also provide real-time feedback, helping employees understand their strengths and areas for improvement as they go through the training modules. This immediate feedback loop is invaluable for both the employee and the organization, as it allows for timely interventions and adjustments to the training program. Companies like Apriorit specialize in enhancing existing LMS platforms with AI capabilities, making them more adaptive and intelligent.

One of the most exciting applications of AI in LMS is predictive analytics. By analyzing historical data and current performance metrics, AI can predict what skills or knowledge an employee might need in the future. This proactive approach allows companies to prepare their workforce for upcoming challenges, ensuring they are always a step ahead.

SelfStudy, a San Francisco-based start-up, offers an AI-powered platform that tailors learning experiences to individual needs. An example of an adaptation is the creation of personalized learning journeys based on analyzing individual learning styles, performance metrics, and career goals. This approach optimizes the training experience by ensuring each employee receives content tailored to their specific needs while enhancing engagement and performance.[8]

OTHER WAYS AI MIGHT REDEFINE HR PRACTICES

Envision a future where AI algorithms scrutinize employee behavior, stress levels, and even biometric data to craft personalized well-being programs. Companies like IBM are at the forefront, utilizing AI to monitor employee satisfaction and mental health, striving for a more balanced and productive work environment.[9]

Conflict in the workplace is inevitable, yet resolving it swiftly and effectively is essential. AI has the potential to analyze communication patterns, employee reviews, and additional data to pinpoint potential conflicts before they worsen. This proactive strategy could enhance workplace harmony and elevate productivity.

Career development often involves a manual, labor-intensive process. AI stands to transform this by continuously evaluating an employee's performance, skills, and career goals, automatically recommending new roles, training programs, or mentorship opportunities. This could streamline career progression, making it more efficient and tailored to individual growth trajectories.

INTEGRATING AI INTO STRATEGIC HR DECISION-MAKING

AI's analytical capabilities extend beyond operational processes, offering valuable insights to inform strategic HR decision-making. By analyzing a multitude of data points—such as workforce demographics, skill gaps, market trends, and organizational goals—AI can support HR leaders in developing data-driven strategies for talent acquisition, development, and deployment.

For example, AI can identify emerging skill gaps within the organization, enabling proactive planning for upskilling and reskilling initiatives. This ensures that the workforce remains agile and equipped to meet evolving business needs.

Additionally, AI can forecast future talent demands based on factors like industry trends, market dynamics, and organizational growth plans. This predictive capability allows HR teams to develop long-term talent acquisition and development strategies, ensuring a continuous pipeline of skilled and engaged employees.

By integrating AI into strategic HR decision-making processes, organizations can proactively address talent challenges, optimize resource allocation, and align their workforce strategies with overall business objectives, fostering a competitive advantage in today's dynamic and rapidly changing business environment.

CASE STUDY: AI IN TALENT ACQUISITION— UNILEVER'S SUCCESS STORY

Unilever, a global consumer goods company, faced significant recruitment challenges with outdated and inefficient processes. It took approximately four to six months to process 250,000 applications to select eight hundred candidates for their Future Leaders Program. This cumbersome process was not suitable for meeting the demands of a growing millennial workforce.

To address these challenges, Unilever introduced a streamlined process consisting of four key stages: simplified application, where candidates submit their details via a brief online form linked to their LinkedIn profiles; gamified assessments, in which candidates engage in twenty-to-thirty-minute games designed to evaluate competencies like problem-solving and emotional intelligence; AI-powered video interviews, where top candidates participate in video interviews analyzed by HireVue's AI technology; and discovery centre experience, where the most promising candidates partake in activities that simulate daily tasks for a hands-on assessment.

AI played a crucial role through HireVue's AI-powered video interviewing platform. Unilever was able to analyze a vast array of data points from candidate interviews to identify those who best matched the company's criteria for potential success. This technology enabled a more predictive assessment of candidates, ensuring that those who progressed were highly likely to succeed and align with Unilever's values and operational demands.

The implementation of these technology-driven solutions yielded significant improvements. With the aid of AI, 50 percent of the candidates screened were extended job offers, indicating higher precision in the selection process. Overall hiring time was reduced from months to just two weeks. Candidate experience improved with over 80 percent positive feedback. Unilever reduced recruiter screening time by 75 percent within the first year.[10]

CHAPTER 15

Building a Better Product: AI Innovation in Engineering

THE AI INNOVATION IN SOFTWARE DEVELOPMENT

Artificial intelligence, especially generative AI, is bringing tremendous innovations to software engineering and development. AI is automating testing, generating new code, finding defects, and accelerating the overall development lifecycle. It's even automating the creation of machine learning models, known as AutoML. This transformation promises to boost engineering productivity and efficiency. No wonder why in a survey done by Retool 48 percent of respondents indicated that writing code was their top AI use case.[1]

AI tools have significantly changed traditional coding and testing methods. GitHub Copilot, for instance, automates the generation of code snippets and complete code blocks, greatly speeding up the development process and allowing developers to focus more on creative design tasks.[2] Tabnine's AI assistant uses generative AI technology to suggest your next lines of code based on millions of open-source code examples. It can also learn from your code patterns to style and provide tailored code completions that match your coding standards and best practices.[3]

In addition to these, generative AI tools like Gemini, Claude, and ChatGPT provide help—from coding and clarifying programming concepts to aiding in problem-solving. Their use is valuable for quickly creating initial code structures, generating standard code patterns, and providing smart coding suggestions.

AI is becoming smarter at analyzing code for errors and potential improvements. It plays a crucial role in maintaining high code standards and ensuring software runs smoothly. In test automation, AI has shown remarkable progress. Tools like Testim and Applitools use AI for creating test scenarios and automating the testing process, which not only improves the quality of the code but also reduces the time and effort involved in manual testing.[4]

Increasingly architects are using AI models to prototype IT system architectures rapidly based on specific requirements. This capability allows for quicker design iterations, leading to faster time-to-market for new systems. In addition to prototyping, AI can significantly aid in modernizing legacy systems. It achieves this by translating outdated codebases into modern programming frameworks, thereby extending the life and usability of older systems. AI-driven tools can analyze existing code structures and automatically suggest or even implement modifications that align with contemporary frameworks and coding standards.

AI's role in debugging and diagnostics involves parsing system logs and performance data to pinpoint and highlight issues. This capability significantly speeds up the process of identifying and resolving software problems. AI not only identifies these issues but also offers suggestions for fixes, streamlining the debugging process.

Additionally, AI can automate the generation of documentation from engineering inputs. This feature is especially beneficial for complex systems where keeping documentation current can be a challenge. By ensuring thorough and up-to-date documentation, AI assists in simplifying future maintenance and development efforts. AI tools in diagnostics can also foresee potential system issues or performance challenges before they escalate, allowing for preemptive maintenance and system enhancement.

Early estimates show generative AI could increase engineering productivity by 20–45 percent per McKinsey.[5] However, quality and accuracy concerns remain. Leading technology companies are pioneering AI integration into development, redefining the future of software engineering.

AI IN PRODUCT R&D

Advanced AI technologies are poised to dramatically reshape the landscape of research and development (R&D). AlphaFold, with its groundbreaking capability in predicting protein structures, has condensed years of complex biological research into days, profoundly impacting fields from pharmacology to environmental science.[6]

Similarly, GNoMe represents a leap forward in material science, enabling the discovery of new materials at an unprecedented pace. Its application in analyzing and predicting material properties transcends traditional boundaries.[7] IBM Research published a study demonstrating how advanced AI algorithms successfully modeled clinical trials to find new uses for existing drugs, particularly for neurological diseases like Parkinson's.[8] Together, these AI-driven tools are not just enhancing the speed but are also enriching the quality and scope of research across diverse fields, reshaping the very fundamentals of innovation and discovery.

Procter & Gamble leverages AI to analyze consumer behavior and preferences, tailoring its R&D efforts to develop products that are more likely to succeed in the market. Vittorio Cretella, the chief information officer at Procter & Gamble, emphasizes the significant impact of modeling and simulation methods in expediting the discovery of new ingredients. These techniques can drastically reduce the time needed, turning months of testing into mere weeks to meet specific functional requirements. Furthermore, he notes the potential of algorithms in processing customer feedback on product modifications. These algorithms can instantly alert R&D engineers and suggest appropriate modifications, enhancing the responsiveness and efficiency of the product development process.[9]

Nvidia is actively leveraging artificial intelligence and machine learning to advance the capabilities of its GPUs.[10] Bill Dally, Nvidia's chief scientist and senior vice president of research, revealed plans to integrate AI and ML into Ada

Lovelace GPUs. They are employing AI and ML to refine various aspects such as power voltage drop monitoring, preemptive error identification through parasitic prediction, automated standard cell migration, and solving complex routing issues.

The application of AI and ML in these areas is expected to markedly improve the speed and efficiency of GPU design processes, potentially reducing the time required for a typical GPU design task from three hours to just three seconds and increasing precision by up to 94 percent. Moreover, NVIDIA's dedication to incorporating AI and ML in hardware development is clearly reflected in their R&D initiatives and public discourse on how these technologies contribute to GPU performance enhancement.

IMPACT OF AI ON DESIGN AND PROTOTYPING

AI algorithms have transformed the field of product design into a dynamic and highly efficient process. Companies like Autodesk are at the forefront, using AI to integrate consumer behavior, market trends, and material properties into their design recommendations. This not only accelerates the design phase but also ensures products are perfectly aligned with consumer needs while minimizing environmental impact.

Autodesk leverages generative design to evaluate countless options within seconds—akin to a high-speed brainstorming session with the best engineers. This method not only produces optimized designs but also significantly reduces the time and resources spent on prototyping. For example, when designing a new drone, generative design evaluated thousands of potential configurations, streamlining the process from conception to final prototype in a fraction of the traditional time.[11]

Following design selection, companies like ANSYS take the helm, utilizing AI to simulate and test prototypes under varied conditions such as fluid dynamics and structural stress. This virtual prototyping, enhanced by AI, mirrors real-world conditions so closely that it drastically reduces the need for physical prototypes, saving further time and materials.[12]

Leveraging its three-decade expertise in 3D printing technology, Materialise employs artificial intelligence to optimize the selection of materials, ensuring high performance while prioritizing sustainability. Their commitment to ecological considerations is reshaping how industries like automotive and aerospace approach material selection, crucial for both performance and environmental stewardship.[13]

Virtual prototyping has also evolved, thanks to AI's capabilities to mimic real-life wear and tear, weather conditions, and even human interaction. This advanced simulation provides invaluable insights, enabling products to be market-ready faster and with greater confidence in their performance and durability.

As AI continues to learn from each simulation, improving with every iteration, the future of prototyping promises unprecedented levels of detail and accuracy. This progress is setting a new standard for innovation and efficiency in product development. Embracing these AI-driven tools is essential for any leader looking to thrive in a rapidly evolving market.

PREDICTIVE MAINTENANCE

A production manager's dream would be production lines never halt unexpectedly, and the quality of the products is consistently top-notch. That dream could be a reality with AI. Predictive maintenance and quality control are two areas where AI is making a substantial impact, ensuring that products not only meet but exceed expectations. Let's explore how AI is dramatically improving these critical aspects of engineering and product development.

The traditional model of maintenance is reactive; you fix things when they break. However, AI enables a shift to predictive maintenance, where algorithms analyze data from machinery to predict when a failure is likely to occur. This allows for timely interventions, reducing downtime and increasing efficiency.

Since the 2017 launch of its open data platform Skywise, Airbus has pioneered several artificial intelligence applications to optimize aircraft maintenance.[14] A keystone application is predictive maintenance, which utilizes machine learning algorithms to determine optimal servicing times for individual components rather than relying on fixed schedules. Seeking to expand its capabilities, Airbus also developed the reliability application to pinpoint technical issues across entire fleets and the health monitoring application to compile real-time aircraft data, detecting equipment events to assist operators with time-sensitive decisions. Through these ongoing efforts, Airbus aims to continuously improve maintenance procedures and operational efficiency.

Quality control is another area where AI is making significant strides. Machine learning algorithms can analyze product data, identify anomalies, and even predict future quality issues before they happen. This ensures that only the highest quality products make it to market. Companies like Qualitas are pioneering AI-driven quality control. Qualitas has developed a visual inspection system called EagleEye, which optimizes its AI models and automates quality control for manufacturing companies.[15]

The system uses computer vision and deep learning to scan products on production lines, identifying defects and quality issues with 100 percent accuracy. This allows manufacturers to catch problems early, reduce wasted materials, and improve overall quality. This enables real-time quality checks rather than waiting for a sample to be manually reviewed. By combining advanced AI with automated optical inspection, Qualitas' solution minimizes costs and maximizes quality across an entire manufacturing workflow.

The future of predictive maintenance and quality control is incredibly promising, focusing on creating better, more reliable products.

FLEXIBLE, SECURE, AND STREAMLINED IT OPS WITH AI

I caught up recently with Farhan Haider, the founder of Aufsite, a leading provider of cloud support services to small businesses. Aufsite's operations predominantly involve coding and task iteration. By retooling to integrate AI, specifically a locally hosted OpenAI GPT instance, the company has significantly enhanced its efficiency. This technology enabled them to automate the creation of complex scripts that previously took hours to develop, now completed in just minutes. Farhan's team meticulously reviews, edits, and rigorously tests these scripts before deployment, ensuring the code is robust and secure. Farhan views AI as crucial to scaling his business and maintaining agility in providing solutions. Although building business relationships remains inherently human, many operational processes, such as task management and IT problem resolution, are increasingly automated, reducing the need for human intervention.

The application of AI in information technology operations is markedly changing the landscape, shifting toward a more flexible, secure, and effective IT landscape. Monitoring and incident management in IT Ops can be significantly improved with AI. Traditional, manual-based methods often miss real-time detection and resolution of issues. AI, adept at analyzing extensive data, excels in pinpointing irregularities and preempting system failures.

Splunk has introduced AI offerings to accelerate detection, investigation, and response across security and observability. It combines automation with human-in-the-loop experiences, allowing organizations to control how AI is applied to their data. For ITOps, it drives faster and more accurate alerting by using historical data and patterns to create dynamic thresholds while detecting and omitting abnormal data points for precise thresholding. For SecOps, it empowers teams with rapid detections by incorporating ML-powered detections to address ongoing security threats and providing access to ML technology, forecasting, and predictive analytics. This enables richer ML-powered insights—ultimately helping to accelerate productivity and lower costs.[16]

Machine learning algorithms enable AI tools not just to identify issues but also to recommend or autonomously enact resolutions. This feature is invaluable in fast-paced, intricate environments where quick, accurate responses are key. IBM's Watson AI, for example, can interpret system logs and performance data to diagnose and remedy problems efficiently. Watson AIOps uses several machine learning models to automatically recognize logs that are reporting service errors in real time.[17]

AI refines IT operations management. It automates routine activities such as patch management and network configuration, allowing IT staff to focus on strategic tasks. ServiceNow incorporates AI in its IT service management solutions, automating various operations and providing AI-derived insights for improved decision-making. AI also significantly contributes to the efficient management of IT resources. By evaluating usage trends and forecasting future needs, AI orchestrates resource distribution, ensuring systems perform optimally while reducing excess.[18]

In IT infrastructure maintenance, AI's predictive analytics are invaluable. Anticipating hardware failures and software glitches, AI facilitates early intervention, decreasing downtime and prolonging IT assets' lifespan. This approach not only conserves resources but also ensures a more dependable IT framework.

AI in IT Ops significantly betters both customer and employee experiences. AI-driven chatbots and virtual assistants, like those created with Microsoft's Azure AI, offer immediate, insightful responses to user inquiries, boosting satisfaction and operational efficiency. For IT teams, AI aids in more effectively managing and solving issues, fostering a more productive work environment.

With cyber threats becoming more intricate, legacy security defenses often fall short. AI-enhanced security systems like Darktrace utilize machine learning to identify unusual activity and potential security breaches, offering a more dynamic defense against evolving digital threats.[19]

CASE STUDY: GE'S AI-POWERED PREDICTIVE MAINTENANCE SOLUTION

General Electric (GE) faced a crucial challenge. Its vast fleet of industrial equipment like wind turbines, jet engines, and power plants required frequent maintenance to prevent costly unplanned downtimes. Traditional preventive maintenance schedules were inefficient, leading to unnecessary upkeep or missed issues that caused disruptions. GE sought a smarter solution to predict and address failures before they occurred.

To tackle this problem, GE implemented an AI and machine learning-powered "Digital Twin" solution for predictive maintenance across its businesses. This innovative approach created virtual models that mirrored GE's physical assets. Sensor data from the equipment was continuously fed into deep learning models that could detect anomalous patterns and deviations from normal operation parameters. By analyzing temperatures, lubricant levels, weather data, power output and more, the AI could accurately predict when a specific component was likely to fail and recommend preemptive actions. Instead of relying on fixed schedules, maintenance could now be scheduled proactively based on the actual condition of each asset.

The results of GE's predictive maintenance solution have been transformative. The AI system has analyzed over 100 million hours of operational data, reducing unplanned downtime by 25 percent and extending the lifespan of assets by 20 percent. Across GE's businesses, this AI-driven approach has delivered annual savings of one billion dollars from increased productivity. In the renewable energy division alone, GE avoided over one thousand unplanned outages in 2020, resulting in $193 million in productivity gains. By harnessing AI to maximize industrial asset uptime, GE has reaffirmed its commitment to driving efficiency and innovation.[20]

CHAPTER 16
AI's Impact on Health-Care Industry

A few years ago, Wendy, a dear friend, found herself in a situation emblematic of the health-care industry's profound challenges. Lying on an operating table, a tumor discovered near an artery, she embodied the agony of uncertainty that plagues many patients. Her medical team, engulfed in a quandary, needed to ascertain if the tumor was malignant to initiate appropriate treatment. Hours ticked by, filled with anxiety and fear, as they awaited the biopsy results—a scenario far too common, especially for those with brain tumors.

Amid such narratives, the health-care industry confronts a labyrinth of challenges. Skyrocketing costs, disparities in health-care access, and an aging global population intensify the need for innovative solutions. The sector struggles to manage and interpret vast data volumes, ranging from patient records to evolving research, all while upholding precision and confidentiality. Furthermore, the scarcity of health-care professionals, especially after COVID and more notably in remote regions, exacerbates these hurdles.

In this intricate landscape, AI emerges as a transformative solution. Its potential is vividly illustrated in the heart of Amsterdam. Researchers at the Amsterdam University Medical Centers and Princess Máxima Center for Pediatric Oncology have developed a pioneering AI system. This system, trained on four neural networks, can swiftly and accurately classify brain tumors using preliminary DNA sequences, offering surgeons newfound confidence in tumor classification.

This advancement has shown tangible results. With the algorithm doctors can identify the tumor type within 20 to 40 minutes.[1] Such efficiency is vital, providing crucial information to brain surgeons at the early stages of operations. If Wendy's doctor had access to this technology, the agonizing wait could have been significantly less.

AI's impact in health care goes beyond patient diagnosis and treatment. It aids in early disease detection, personalizing treatment plans, and managing administrative tasks, thus allowing health-care professionals to focus more on patient care. Its effectiveness in handling health crises, such as the COVID-19 pandemic, highlights its rapid response capabilities. This chapter explores AI's multifaceted role in health care—from improving medical diagnoses to reinventing hospital operations—ultimately making health care more accessible and efficient.

AI IN MEDICAL DIAGNOSIS AND TREATMENT

The landscape of medical diagnosis is undergoing a remarkable change, spearheaded by the integration of artificial intelligence. Traditional diagnostic methods, while effective, often involve a degree of uncertainty and are prone to human error. AI, with its advanced algorithms and machine learning capabilities, brings a new level of precision and speed to the diagnostic process.

AI systems excel in analyzing intricate medical images, such as X-rays, CT scans, and MRIs, with a level of detail and consistency that can sometimes surpass human capabilities. For instance, AI algorithms have demonstrated exceptional accuracy in identifying early signs of diseases like cancer, often detecting malignancies that even the most skilled radiologists may initially overlook. A remarkable study conducted by the Mass General Cancer Center and the Massachusetts Institute of Technology yielded an AI tool named Sybil, which accurately predicts whether an individual will develop lung cancer within the next year with an astonishing accuracy rate of 86 percent to 94 percent.[2] This groundbreaking tool can detect early indicators of lung cancer years before they would become visible on a CT scan, showcasing the immense potential of AI in advancing early disease detection and intervention.

The impact of AI in diagnostics is not just about accuracy; it's also about accessibility. In areas where medical experts are scarce, AI-powered diagnostic tools can provide critical support, offering reliable evaluations that guide treatment decisions. This democratization of diagnostic capabilities is a significant stride toward equitable health care.

Cancer Detection and Treatment: One of the most groundbreaking applications of AI is oncology. Google's

DeepMind developed an AI system that can match the performance of human radiologists in detecting breast cancer from mammograms. This AI model reduces false positives by 1.2 percent to 5.7 percent and false negatives by 2.7 percent to 9.4 percent, offering a more reliable diagnosis.[3]

Moreover, AI-driven platforms are enabling personalized cancer treatment by analyzing patients' genetic makeup and cancer's molecular structure, recommending tailored strategies likelier to be effective, though yet to be clinically validated.

Deep learning, especially convolutional neural networks (CNNs), effectively detects cancer in medical images like mammograms and CT scans. CNNs learn patterns from large datasets of labeled images, distinguishing cancerous from noncancerous cases. Transfer learning fine-tunes pre-trained CNNs on specific medical data, reducing data needs. Research shows CNNs can achieve human-level accuracy in detecting breast cancer from mammograms.[4]

At a volunteer event, I ran into Seymour Duncker, the founder of Mindscale.ai. He shed light on some alarming statistics regarding breast cancer. Shockingly, over 50 percent of women miss critical follow-up breast cancer screening appointments. Furthermore, a staggering 22 percent of breast cancer cases are detected at an advanced stage, emphasizing the urgency for improved screening and early detection measures. Seymour's team is leveraging AI for early cancer detection and risk mitigation.

IBM's Watson Health is another example. It assists radiologists by identifying key markers in medical images that could indicate the presence of diseases like cancer. Watson Health has been trained on a vast dataset of medical images, enabling it to spot even the most subtle irregularities that could be early signs of cancer. This not only speeds up the diagnostic process but also improves its accuracy, potentially saving lives.[5]

Cardiovascular Diseases: AI algorithms are optimizing the detection and management of heart diseases by analyzing large datasets and identifying risk factors often missed by humans. For example, an AI tool developed by researchers at the University of Nottingham predicts heart attack risk more accurately than traditional methods. This tool, which uses machine learning models like neural networks and random forests, provides nuanced risk assessments by analyzing patient data, including medical records and lifestyle factors. These models are trained on large datasets to identify patterns and subtle connections. Techniques like feature selection help the AI identify the most predictive risk factors, enabling the tool to improve treatment for up to 45 percent of patients and potentially save thousands of lives by identifying those at risk of a heart attack who might otherwise be missed.[6]

Neurological Disorders: In neurology, AI is assisting in the early detection and treatment of conditions such as Alzheimer's and Parkinson's disease. Researchers at the University of California, San Francisco, used machine learning to identify a specific pattern of brain activity in EEG tests that predicts seizures in epilepsy patients. This innovative tool facilitates proactive disease management, markedly enhancing the quality of life for these individuals.[7] Continued research and development are essential to refine the accuracy and dependability of these seizure prediction models, promising even greater advancements in patient care.

Diabetic Retinopathy: The FDA-approved IDx-DR system is an AI-based diagnostic tool that autonomously analyzes retinal images to detect diabetic retinopathy, a leading cause of blindness. The device correctly identified the presence of more than mild diabetic retinopathy 87 percent of the time and correctly identified those who did not have more than mild diabetic retinopathy 90 percent of the time, thereby enabling early detection and treatment, preventing the progression of the disease.[8]

Remote Diagnosis and Telemedicine: AI is playing a crucial role in telemedicine, particularly in remote and underserved areas. AI-powered diagnostic apps allow patients to receive medical assessments from their homes, broadening access to health care. These apps analyze symptoms and medical history, offering preliminary diagnoses and recommendations, thus bridging the gap in health-care accessibility.

The integration of AI in medical diagnosis and treatment is a testament to the technology's potential in health care. By enhancing diagnostic accuracy, personalizing treatment plans, and democratizing access to health-care services, AI has become a catalyst for a new era in medical care.

REVOLUTIONIZING DRUG DISCOVERY AND DEVELOPMENT

In September 2021, amid the turmoil of the COVID-19 pandemic, my wife and I made a trip to Chicago to visit her father, who lived in an elder-care facility. It was a bittersweet reunion. We sat outdoors six feet from him and were delighted by his cheerful demeanor. Little did we know, it would be our last meeting.

As Thanksgiving approached, my wife became anxious. Her dad usually calls her a few times a week, but she didn't get a call from him in days. She reached out to the nursing home multiple times. Eventually, we received the dreaded call: He was ill with a fever, and COVID had infiltrated the facility. Due to the overwhelming number of cases in Chicago hospitals at that time and the dire shortage of available beds, he was briefly assessed and sent back to the elder-care facility under the assumption he could recuperate there. Tragically, just five days later, his condition deteriorated, necessitating urgent hospitalization. By then, it was too late; he suffered multiple strokes and entered a vegetative state from which he would not recover.

Heartbreakingly, he was just three weeks shy of receiving the COVID vaccine—three weeks from a shot at protection. The protracted process of vaccine development, clinical trials, and regulatory approvals, which typically ensures safety and efficacy, in this instance, spelled a devastating loss for our family and millions of others.

This poignant experience underscores the urgent need for innovation in how we develop and deploy medical interventions. AI holds the promise to expedite drug discovery and development, potentially shortening these lengthy cycles and delivering crucial vaccines and treatments more swiftly. As we explore this topic, we consider how leveraging AI could prevent countless similar tragedies by ensuring life-saving measures reach those in need with unprecedented speed.

The journey of drug discovery and development is a lengthy and costly process. That will change with AI's ability to analyze vast datasets rapidly and uncover hidden patterns and identifying potential drug candidates and therapeutic targets. This computational power significantly accelerates the initial stages of drug discovery, reducing the time and resources traditionally required.

AI algorithms can simulate and predict how different compounds will interact with biological targets, enabling researchers to identify promising molecules more quickly. This approach is particularly useful in understanding complex diseases where multiple factors are at play. By analyzing genetic, environmental, and lifestyle data, AI can pinpoint specific pathways and targets for intervention, tailoring drugs to address the root causes of diseases.

Moreover, AI contributes to a more efficient screening process. High-throughput screening, where thousands of compounds are tested for activity against a biological target, is enhanced by AI algorithms that can predict the most likely candidates for success, thereby streamlining the selection process.

Here are some ways AI is helping with the discovery and development of drugs.

AlphaFold and Protein Folding: A remarkable example of AI's impact in pharmaceuticals is DeepMind's AlphaFold. In 2020, AlphaFold made a significant breakthrough in predicting protein structures, a challenge that had puzzled scientists for decades. By accurately predicting the 3D shape of proteins from their amino acid sequences, AlphaFold has opened new frontiers in drug development. It utilizes an AI architecture inspired by transformers and self-attention mechanisms, trained on a vast dataset of known proteins with known 3D structures mapped to their corresponding 1D sequences. During inference, AlphaFold uses the 1D sequence to predict the 3D structure with remarkable accuracy of its predictions to experimentally derived structures at the highest level of molecular geometry. This breakthrough demonstrates AI's potential to accelerate biological research by rapidly inferring 3D protein structures, critical for understanding their functions and developing new therapeutics.[9]

Oncology Drug Development: AI is playing a crucial role in developing new cancer therapies. For instance, Atomwise uses AI algorithms to analyze molecular structures and predict how they will interact with cancer targets. This has led to the identification of novel compounds that could potentially treat hard-to-target cancers.[10]

Neurodegenerative Diseases: In the field of neurodegenerative diseases, AI is aiding in the discovery of drugs for conditions like Alzheimer's and Parkinson's. A research team from Nagoya University in Japan has developed a groundbreaking AI technology, named "in silico FOCUS," for analyzing cell images. This AI uses machine learning to predict the therapeutic effects of drugs on neurodegenerative disorders, such as Kennedy disease. Traditional treatments for these diseases often have severe side effects, and the lack of effective screening technologies has hindered the search for safer treatments. AI platforms are offering hope in areas where traditional drug discovery has struggled.[11]

Custom Treatments for Rare Diseases: AI's transformative role in drug discovery is exemplified in the fight against rare diseases like Fragile X syndrome, the most common inherited cause of autism. Right now, we don't have effective treatments for it. Leveraging its ability to analyze vast datasets, companies like Healx utilize AI for omic-based drug matching, identifying promising candidates at a pace and precision unattainable by traditional methods.

This approach led to the discovery of HLX-0201, illustrating how AI accelerates the discovery process by efficiently identifying potential treatment-disease links, a task that would otherwise take humans centuries to accomplish.[12]

Predicting Drug Interactions: Researchers from the Gwangju Institute of Science and Technology in South Korea have developed a deep learning model that predicts adverse drug-drug interactions (DDIs) based on their effects on gene expression. The deep learning model, comprising a feature generation and a DDI prediction model, enhances drug safety by predicting a drug's impact on gene expression and forecasting side effects from drug combinations. This innovative approach aids in defining drug usage during development and has the potential to dramatically accelerate drug development processes, ensuring safer patient outcomes and efficiency. The model's ability to predict DDIs for both approved and novel compounds addresses the risks of polypharmacy, making it a significant advancement in drug safety monitoring.[13]

ENHANCING PATIENT CARE AND EXPERIENCE

The integration of artificial intelligence in health care is significantly improving the quality and customization of patient care, leading to enhanced outcomes and patient satisfaction. AI's advanced data analysis and predictive capabilities enable health-care providers to offer more focused and individualized care.

Sophisticated AI technology has brought new precision to patient monitoring, providing accurate and continuous health condition tracking. Companies like Fitbit and Apple have pioneered in this domain with wearable devices. Wearable devices, equipped with AI algorithms, monitor vital health statistics such as heart rate and blood glucose levels in real time, offering immediate insights to both patients and health-care professionals.[14] This technology is especially valuable in managing chronic diseases, where early detection of changes can be critical in preventing severe complications.

In hospital settings, AI systems from companies like Philips are used for vigilant monitoring of patients in post-surgical recovery or intensive care. These systems can detect minute changes in a patient's status, which might signal the need for urgent medical attention, often before these issues become apparent to the clinical staff.[15]

AI's role in tailoring treatment plans to individual patients is reshaping how health care is delivered. Analyzing extensive data including medical history, genetic information, and current health indicators, AI can recommend the most effective treatment strategies. This custom approach not only boosts the effectiveness of treatments but also minimizes potential side effects.

Leveraging training from oncologists at Memorial Sloan Kettering, IBM's Watson for Oncology uses natural language processing (NLP) and machine learning to analyze individual patient data, such as test results, tumor details, and genetic mutations. The NLP component structures clinical literature, treatment guidelines, and patient records, extracting insights like patient characteristics and cancer staging. Machine learning, trained on historical patient datasets, estimates probable treatment outcomes. Watson rapidly combs through millions of pages of medical literature using probabilistic algorithms to identify the most relevant, evidence-based treatment options tailored to each patient's unique cancer profile. Studies show Watson's recommendations achieve up to 96 percent concordance with physician recommendations across major cancer types, making it a powerful decision support tool in cancer care.[16]

In mental health, AI-driven platforms are offering customized therapy and support. Woebot Health leverages AI to offer accessible, convenient mental health support by employing a rules-based system that mimics cognitive behavioral therapy (CBT). Their chatbot is trained on extensive, specialized datasets to recognize words, phrases, and even emojis associated with dysfunctional thoughts, allowing it to challenge negative thinking patterns. Woebot provides immediate guidance and tools to manage symptoms like depression and anxiety. "We know the majority of people who need care are not getting it. There's never been a greater need, and the tools available have never been as sophisticated as they are now," said Alison Darcy, the founder and president of Woebot Health.[17] By being available on smartphones twenty-four-seven, it addresses the critical shortage of therapists and stigma associated with seeking therapy, helping to bridge the gap for those unable to access traditional mental health care.[18]

AI IN HOSPITAL OPERATIONS AND ADMINISTRATION

AI opens the door for a new level of efficiency in administrative tasks. For example, Qventus utilizes AI to manage various hospital operations, combining machine learning and behavioral science to automate care operations. This integration results in reduced inpatient and emergency department lengths of stay as well as fewer patients leaving without being seen.[19] AI's predictive capabilities are crucial in managing hospital resources. By analyzing patient admission trends and historical data, AI can forecast bed availability and staffing needs, as seen in Qventus's platform, ensuring efficient use of resources. AI-driven chatbots, such as those employed by GYANT, handle appointment bookings and patient queries, reducing the workload on administrative staff and minimizing human error.[20]

Corti's AI system exemplifies the impact of AI in emergency response, assisting call dispatchers by analyzing speech and sounds to recognize urgent conditions like cardiac arrest. This not only aids in providing accurate diagnoses but also enhances the overall effectiveness of emergency health-care delivery.[21] By improving the accuracy of health data through precision diagnosis and clinical decision support, Enlitic aids in creating more efficient workflows and better patient outcomes.[22]

TELEMEDICINE AND AI-ENHANCED REMOTE HEALTH CARE

The incorporation of AI into telemedicine platforms has dramatically broadened health-care access, especially in remote and underserved areas. AI algorithms augment telemedicine platforms by offering tools for AI-driven diagnostics, symptom analysis, and health monitoring, allowing patients to receive initial assessments and ongoing health monitoring from their homes.

Babylon Health is indeed leveraging AI to provide remote medical consultations. Their AI-powered chatbot evaluates symptoms and offers medical advice or referrals, a method that has proven crucial in triaging patients and ensuring they receive timely and appropriate care.[23]

Similarly, Biofourmis is employing AI for remote monitoring, particularly for managing chronic diseases. Their platform analyzes data from wearable devices to monitor patient health in real time, providing insights that enable proactive management of chronic conditions.[24] AI models analyze continuous, multimodal data from wearables to identify deviations from each patient's baseline that may indicate a risk of an adverse event, enabling preventive interventions before the condition deteriorates.[25]

AI is also making significant strides in remote mental health support. Platforms like Talkspace utilize AI to analyze speech and text for mental health assessment and to deliver cognitive behavioral therapy techniques. These platforms have demonstrated effectiveness in expanding access to mental health care and offering personalized therapeutic experiences.[26]

Moreover, AI-powered telemedicine apps are analyzing health data to offer personalized health and wellness advice. These apps offer not only medical recommendations but also lifestyle and nutritional guidance tailored to the individual's health status and goals. These advancements in AI are transforming the health-care industry by making health care more accessible and personalized.

AI IN MEDICAL TRAINING AND EDUCATION

AI is reshaping medical training and education by introducing advanced methodologies that enhance learning experiences, tailor educational content, and simulate real-world medical scenarios. By creating interactive, responsive, and personalized learning experiences, AI is ensuring medical professionals are better equipped to handle the complexities of modern health care.

AI integrates with VR and AR to create detailed simulations of medical scenarios. These simulations are built using vast datasets of medical cases, enabling them to replicate a wide range of clinical situations. AI algorithms adjust these scenarios in real time based on the learner's interactions, providing a responsive environment that mimics real-life decision-making and procedures. In addition, AI systems analyze individual learning behaviors, including the pace of learning and areas of strength or weakness. Using this data, AI tailors educational content to match each learner's profile, adjusting the complexity and focus of material to optimize the learning experience.

By evaluating responses to quizzes, simulations, and interactive content, AI identifies areas where learners need additional focus. The platform then adjusts the curriculum, presenting new or revised material that targets these specific areas. For clinical research training, AI tools sift through extensive research publications and clinical trial data. They employ natural language processing to extract key findings and trends, presenting these in an accessible format for learners. This keeps medical professionals updated with the latest research, enhancing their understanding of evolving medical practices.

AI's contribution to training health-care professionals is multifaceted, involving several innovative approaches. It creates virtual patients by synthesizing medical histories, symptoms, and potential complications based on real patient data. Learners interact with these virtual patients, making diagnostic and treatment decisions. AI evaluates these decisions against established medical guidelines and patient outcomes, providing immediate feedback and suggestions for improvement.

In surgical training, AI algorithms analyze video recordings of surgical procedures using image recognition and machine learning. These algorithms assess surgical techniques against best practices, offering objective feedback on the precision, timing, and efficiency of each movement and decision the trainee makes. This feedback is crucial for refining surgical skills in a controlled, repeatable environment.

Case Western Reserve University's School of Medicine developed the HoloAnatomy mixed-reality software for the Microsoft HoloLens, which allowed students to learn anatomy remotely during the COVID-19 pandemic.[27] Osmosis is a medical education platform that offers a vast library of resources, including videos, flashcards, and case questions, to help future health professionals master vital health science information.[28] GIBLIB provides medical lectures and surgical procedures from top medical centers, as well as VR-based medical education content, such as accredited VR continuing medical education courses, that allow learners to gain knowledge from leading doctors in an immersive format.[29]

ETHICAL CONSIDERATIONS AND CHALLENGES

While you read about the ethical considerations for AI in chapter 9, the criticality of them for health care cannot be overemphasized. The integration of AI in health care brings with it a myriad of ethical considerations. One of the primary concerns is ensuring AI applications are developed and used in a manner that is fair, transparent, and nondiscriminatory. We need to address potential biases in AI algorithms that could lead to unequal treatment of different patient groups. Additionally, there is a need for ethical decision-making frameworks that guide AI's role in patient diagnosis and treatment, ensuring these technologies complement rather than supplant human judgment and expertise.

Privacy and data security are paramount in health-care AI applications due to the sensitive nature of medical data. Ensuring the confidentiality and integrity of patient data requires robust security measures, especially as AI systems often require large datasets to function effectively.

Security involves implementing strong data encryption, access controls, and regular security audits. Furthermore, there is a need for clear regulations and guidelines on data usage, consent, and sharing to ensure patient data is used responsibly and ethically. Navigating these privacy concerns requires a collaborative effort among technologists, health-care providers, and policymakers to develop standards that protect patient information while enabling the benefits of AI in health care.

CASE STUDY: OPTIMIZING NURSE ASSIGNMENTS

A Fortune 500 health-care company faced significant challenges in managing nurse assignments across its 186 hospitals. The manual processes in place led to inefficient workload distribution among nurses, which not only caused inconsistencies in task assignments but also resulted in substantial time wastage, affecting overall operational efficiency.

To address these challenges, the company collaborated with A.Team, a company known for its skilled product builders. Together, they initiated an idea accelerator project rooted in design thinking principles to devise a solution. The outcome was the development of a cutting-edge application designed to optimize the nurse assignment process. Utilizing artificial intelligence, this application was equipped to analyze complex data related to workload intensity, capacity, and unit capabilities, thereby enabling a more balanced and efficient distribution of tasks among nurses.

The implementation of this AI-driven application significantly transformed the operational dynamics at the company. By automating the nurse assignment process, the system not only reduced the time previously spent on manual scheduling but also achieved a more equitable task distribution. This led to enhanced operational efficiency and improved quality of care, as nurses could dedicate more time to patient care rather than administrative duties. The case illustrates the potential of AI to streamline health-care operations and highlights the importance of technological partnerships in innovating health-care management practices.[30]

CHAPTER 17

AI's Transformative Influence: From Vineyards to Outer Space

Let's take a detour from the expected and explore the unlikely intersection of AI and...winemaking.

FROM ALGORITHMS TO VINEYARDS: AI RESHAPING WINEMAKING

During my engaging conversation with Professor Angelo Camillo, we uncovered the intriguing ways AI is transforming the traditional art of winemaking. From the vineyard to the barrel room, AI is fermenting a reform in this age-old craft.

At its core, AI empowers data driven decision-making throughout the winemaking process. Sensor networks and drone imagery provide a level of monitoring for soil health, weather patterns, and grape maturation that was unimaginable before. This real-time data intelligence allows winemakers to make precisely targeted interventions—optimizing irrigation, maximizing yields, and even predicting disease outbreaks before they occur.

But AI's prowess extends far beyond just the vineyards. In the production facility itself, artificial intelligence is automating and fine-tuning the most complex processes. Machine learning systems can continuously monitor and control every nuance of fermentation, ensuring unparalleled consistency from batch to batch. Computer vision can inspect bottles with microscopic precision, guaranteeing quality levels that sidestep human fallibility.

Remarkably, AI isn't just enhancing efficiency. It's future-proofing winemaking for sustainability too. By comprehensively modeling resource usage, AI can minimize waste, optimize water irrigation schedules, and promote eco-friendly practices at every step of production. The environmental impact is being uncorked.

Yet AI's benefits pour out far beyond the cellar doors as well. Machine learning can demystify the oft-intimidating world of wine by guiding inexperienced consumers through their first tastings based on personal preferences. Recommendation engines can even predict your own "perfect pairing" based on individual taste profiles and purchase histories.

As Professor Camillo and I discussed, AI is both hoisting and swirling the wine industry into a new era of refinement and delight. With artificial intelligence as winemaking's latest blend, the future of this timeless craft has never looked so well-rounded and full-bodied.

Adding to this narrative of innovation is the story of Palmaz Vineyards. COO Christian Palmaz showed me how they are leveraging AI to elevate the human craft of winemaking. At the core of this technological innovation is an AI-powered system called FILCS that Christian developed in 2014 and improved over time. The AI system is designed to augment human winemakers by taking over the tedious monitoring of fermentation curves and precise temperature control adjustments for each of their 24 individual fermentation lots.

"FILCS doesn't replace the winemaker. It allows them to be a better winemaker," Christian explained. By using a vibrating fork to measure density changes and machine learning to model the ideal fermentation conditions, FILCS can continuously fine-tune the temperature zones of each tank with split-second precision impossible for human winemakers alone.

This frees up the Palmaz winemaking team to focus on the qualitative assessments of color, aroma, texture, and taste evolutions happening throughout fermentation—the artistic aspects that drive their blending decisions which AI cannot replicate.

A pivotal moment highlighted FILCS's transformative impact. Palmaz shared with me that the once-common issue of fermenters getting "stuck" during fermentation—a challenge that historically surfaced every couple of years—has been virtually eliminated since FILCS's deployment. Its ability to predict potential stuck fermentation days in advance and make real-time adjustments to avoid them has been game changing.

FILCS was trained on over a petabyte of Palmaz's historical fermentation data, including human feedback on what conditions produced ideal results. Armed with this knowledge, it can provide a level of quantitative process control and predictive capability impossible for a human winemaker alone. The combination of human artistry and machine optimization is taking winemaking at Palmaz to new heights of quality and consistency.

AI'S ROLE IN ENHANCING AGRICULTURE

AI is carving a significant niche in agriculture, presenting diverse applications that span the entire agricultural value chain. Its role is increasingly pivotal in meeting the sustainability and productivity challenges of feeding a rapidly growing global population.

In agriculture, we see a shift to precision agriculture—a modern farming strategy that leverages technology to improve efficiency, profitability, and sustainability in agricultural production. Machine learning is modernizing crop monitoring by analyzing data from satellites, drones, and ground sensors.

Artificial intelligence, particularly deep learning models, has demonstrated exceptional proficiency in identifying diseases, pest patterns, and nutritional deficiencies in crops like ginger with accuracies up to 99 percent. These AI-driven systems utilize extensive real-field image datasets to enable early detection, facilitating timely interventions that enhance crop management and yield. The integration of AI in agriculture promises significant advancements in sustainability and efficiency, significantly altering traditional farming practices by turning vast data into actionable insights.[1]

Taranis provides farmers with innovative and accurate precision agriculture solutions, powered by AI and machine learning, to optimize crop yields and streamline farm operations.[2] Agremo leverages AI to enhance precision farming by analyzing complex data from drone-captured images to increase productivity and streamline plant monitoring throughout the growing season. By employing sophisticated AI algorithms, Agremo enables easier and more accurate plant counting, monitors crop development, and timely detects plant stress, allowing for immediate intervention. This comprehensive analysis helps agricultural professionals to optimize their operations and yield better outcomes from their fields.[3]

Incorporating AI in robotics is transforming farm labor tasks such as harvesting and weeding. An example is Blue River Technology's "see and spray" machinery, which leverages computer vision to identify individual weeds for precise herbicide application. This technology significantly reduces the environmental footprint and costs associated with traditional, less-targeted spraying methods.[4]

AI is instrumental in aggregating and analyzing diverse datasets— including weather patterns, market trends, and production data—to provide customized recommendations for each farm. Platforms like Agriwebb utilize AI to streamline livestock management, optimize planting, and enhance yield forecasts. This approach results in more efficient farm management tailored to the specific needs and conditions of each agricultural operation.[5]

Exploring AI's potential in genetic research, companies like Benson Hill Biosystems are leading seed innovation utilizing proprietary genetics. By analyzing genetic sequences, AI can accelerate the creation of crops that are more resilient, nutritious, and productive, offering a cutting-edge solution to meet the demands of a growing population.[6]

AI IN RETAIL

In the retail and consumer packaged goods (CPG) industries, AI introduces significant advancements, altering traditional approaches to business operations and customer engagement. Explore the art of possibility. As you walk into a store, your smartphone shows personalized recommendations. Or envision supply chains operating with such efficiency that out-of-stock scenarios are eliminated.

Leveraging data analysis, AI crafts individualized shopping experiences. It sifts through customer data, like purchase history and online activity, to suggest products that align with individual preferences, moving beyond generic recommendations. This approach, used effectively by Amazon, not only augments the shopping experience but also boosts sales.[7] In physical stores, technologies like AI-enabled smart mirrors offer interactive experiences, combining the convenience of digital insights with the tangibility of in-store shopping.[8]

In supply chain management, AI applications utilize predictive analytics for precise demand forecasting and inventory control. For instance, Walmart employs AI to predict required stock levels, reducing overstock and stockouts. AI tools also provide real-time responses to supply chain disruptions, ensuring consistent product availability.[9]

AI analyzes comprehensive customer data to extract insights for targeted marketing. Retailers use AI to segment customers and customize marketing messages, thereby enhancing campaign effectiveness.[10] AI's sentiment analysis tools also process social media and review data, offering direct feedback for product improvement and brand perception.

In retail, AI helps to streamline operations through technologies like autonomous checkout systems. These systems reduce wait times and enhance the shopping experience. Similarly, AI-driven robotics are deployed for efficient product packaging and sorting, enhancing productivity and reducing dependence on manual labor.

Additionally, AI enables retailers to implement dynamic pricing, adjusting prices based on factors such as market demand and competition. Best Buy, for example, uses these AI models to adapt prices in real time to market conditions while optimizing profit margins.[11] AI aids in developing products aligned with current market trends and consumer preferences. It predicts emerging trends, allowing companies to design products that meet evolving consumer needs, a step ahead of traditional market research methods.

CASE STUDY: BOOSTING RETAIL SALES AND SECURITY WITH VIDEO ANALYTICS

Bata, a leading global footwear retailer, aimed to improve in-store sales, operations, and security across its five-thousand-plus stores. Specifically, Bata wanted to increase conversion ratios, achieve better efficiency in campaigns and staff operations, ensure compliance with Standard Operating Procedures (SOP), and reduce theft and perimeter breaches. To address these challenges, Bata partnered with Agrex.ai to implement an AI-powered video analytics solution leveraging computer vision and deep learning.

The unified platform analyzed real-time video feeds from existing security cameras to provide predictive sales analytics, audience segmentation, smart conversion tracking and security monitoring. Key features included measuring conversion ratios and customer journeys, forecasting revenue through predictive modeling, detecting theft attempts with 95 percent accuracy, and analyzing customer demographics and peak footfall timings. The scalable solution delivered real-time actionable insights across hundreds of retail outlets. Within months of implementation, Bata witnessed a double-digit percentage increase in conversion ratios. The AI-powered platform also substantially improved marketing efficiency, staff productivity, customer satisfaction, revenue per square foot and loss prevention across stores. The strategic use of video analytics helped optimize operations, drive sales, elevate security and deliver strong ROI—paving the path for continued AI innovation across Bata's retail empire.[12]

LEVERAGING AI IN SPACE INDUSTRY

Space exploration has always been a frontier that challenges our limits and expands our understanding of the universe. The advent of AI is poised to dramatically alter how we explore and interact with outer space. While AI's role in space is still in its infancy, its potential applications are vast and varied.

One key area is using machine learning models to analyze spectral data from satellites or rovers to identify minerals and other valuable resources, akin to a "virtual prospector." NASA's Curiosity Rover on Mars uses AI to study soil samples and identify minerals that could support life.[13]

AI is also being used to optimize supply chain logistics for spacecraft. By analyzing satellite images, AI can determine the most efficient routes for spacecraft to collect supplies from multiple space stations.[14]

Inventory management is another promising application, with AI predicting the lifespan of essential supplies and equipment to assist with timely resupply missions. Automated navigation, powered by AI algorithms, can help spacecraft traverse asteroid belts or land on distant celestial bodies by making real-time calculations too complex for humans.[15]

Perhaps one of the most exciting uses is employing AI to sift through massive amounts of data from telescopes and sensors. Machine learning algorithms can identify patterns or anomalies that could indicate new celestial discoveries. The Kepler Space Telescope uses this technique to discover previously unknown exoplanets.[16]

CASE STUDY: FIRST GREENFIELD MINERAL DEPOSIT DISCOVERY USING AI

Earth AI, an innovative mining explorer, has made a groundbreaking discovery in New South Wales, Australia, by locating a previously unmined greenfield molybdenum deposit using AI. This marks the first instance of AI being utilized to uncover a greenfield deposit. The AI system, named Genesis, achieved this feat by processing an extensive array of geological data, including rock types, structures, and geochemistry. Genesis was specifically trained to detect patterns indicative of mineralization. Following the AI's identification of a potential site, Earth AI's geological team conducted on-site investigations, confirming the presence of the molybdenum deposit.

The methodology behind Earth AI's success lies in the Genesis system's use of advanced machine learning algorithms. These algorithms are trained on vast datasets to model the subsurface, allowing the system to identify anomalies and potential mineral deposits. For instance, Genesis can pinpoint areas with unusually high concentrations of certain minerals, suggesting the presence of a deposit. This AI-driven approach is not only faster than traditional human analysis but also capable of uncovering deposits in challenging locations, such as under deep cover or in remote areas.[17]

The implications of this discovery are significant for the mining industry. AI's ability to expedite and enhance the accuracy of mineral exploration presents a transformative opportunity. It can lead to quicker identification of new deposits, discovery of deposits that might be missed by conventional methods, and a reduction in exploration costs through automation of various tasks. Earth AI's achievement in discovering the greenfield molybdenum deposit exemplifies the significant potential of AI in mining, promising lower costs and increased production efficiency.

AI IN DEFENSE: FORCE MULTIPLIER FOR NATIONAL SECURITY

The transformative impact of AI extends far beyond commercial applications like winemaking and retail. AI is emerging as a critical force multiplier in national defense and intelligence operations.

In my interview, Mandy Long, CEO of BigBear.ai, illuminated the vital role AI plays in enhancing maritime security for the US Navy. Their Arcas solution harnesses the power of AI/ML forecasting and computer vision to gather, interpret, and facilitate real-time responses to potential threats in vast maritime environments that have historically proved challenging to monitor.

Arcas leverages advanced AI to identify risks and provide analysts and decision-makers with situational awareness, enabling autonomous fleet operations while reducing the need to deploy human assets for intelligence gathering in contested waters. This AI-driven capability is a game-changer for maritime domain awareness and force protection.

The US Department of Defense has unveiled a new strategy emphasizing the rapid delivery of cutting-edge capabilities, including AI solutions from companies like BigBear.ai. This strategy prioritizes strengthening America's supply chains, fostering international cooperation to safeguard critical infrastructure, and bolstering the nation's defenses against adversarial threats.

As Mandy states, "There is no other option but to leverage advanced technologies like AI in the mission to protect our nation. We need to respond swiftly to threats and employ these technologies to prevent further escalation."

AI is proving indispensable in providing a strategic force-multiplying advantage, enabling the US military and intelligence agencies to make well-informed, data-driven decisions that enhance national security while minimizing risks to human personnel. As AI capabilities continue to advance, their integration into defense operations will only intensify, shaping a future where human analysts and AI systems collaborate seamlessly to safeguard America's interests.

The applications of AI are only limited by our imagination. From AI-driven journalism to automated legal consulting, the possibilities are endless. As AI technology continues to mature, we can expect to see its impact in even the most unexpected places.

The preceding chapters have showcased AI's transformative power across a kaleidoscope of domains. We have witnessed how it is reshaping finance, enhancing customer experience, and driving marketing innovations. AI is optimizing operations—from engineering processes to health-care delivery to agriculture. By thoughtfully embracing this powerful technology, we can open doors to unprecedented opportunities that elevate innovation and progress across every sphere of human endeavor.

SECTION IV

THE HORIZON: PREPARING FOR THE FUTURE OF AI

CHAPTER 18

The AI Maturity Model: Stages of AI Adoption

> *"The AI transformation will be unlike anything we've experienced before in business. But with the right strategy and commitment, it can catalyze your organization's ascent to new heights of innovation and impact."* —Fei-Fei Li

The path to AI adoption is not linear but rather resembles an ascending spiral with organizations progressing through various stages of maturity. Understanding where your organization stands on this AI maturity curve is pivotal, serving as a compass for navigating the exhilarating yet complex journey ahead.

Just like climbing a mountain, realizing the full potential of AI requires grit, perseverance, and a thoughtful approach. You must assess the terrain, equip yourself with the right tools, and then progress gradually through base camp to summit.

AI MATURITY MODEL

The AI Maturity Model is a framework that helps organizations identify their current capabilities and readiness in adopting AI technologies. Think of it as a GPS for your AI journey. The AI maturity model outlined here distills the collective experience of industry leaders who have successfully harnessed the transformative power of AI. Their journeys hold invaluable lessons for organizations aiming to follow suit.

AWARENESS

In the awareness stage, organizations are just scratching the surface of what artificial intelligence can offer. The primary characteristic of this stage is a limited understanding of AI's scope and potential. Conversations around AI at this point are often rife with misconceptions or overly simplified views. A budding curiosity among employees and leaders typically stems from industry buzz, competitor activities, or evolving customer demands.

At this juncture, typically no dedicated team or individual is tasked with steering AI initiatives. Any AI-related activities are usually ad-hoc and may be considered "side responsibilities" for interested employees.

Financially speaking, the organization has not yet committed any significant resources to AI. Budgets, if they exist, are minimal and usually restricted to basic research or educational materials. Occasionally, you might find isolated experiments or pilot projects related to AI, but these lack a clear business objective or strategy.

So, what should organizations at the awareness stage focus on? First and foremost, education is key. Internal workshops can be invaluable for educating employees about the basics and potential of AI. These workshops should be complemented by regular briefings for the leadership team, keeping them abreast of AI trends and their potential business impact. Competitive analysis can also offer valuable insights, serving as a wake-up call for organizations that are lagging.

At this stage, it may also be beneficial to consult with AI experts for an initial assessment and to provide actionable recommendations. Finally, even a small budget allocation for AI education and initial exploration can go a long way. This could be used for activities ranging from attending AI conferences to hiring temporary consultants.

Based on my discussions with many CXOs and board members, most companies across the spectrum of industries are at this stage as of 2024.

EXPLORATION

In the exploration stage, organizations have moved beyond mere awareness and are actively investigating how AI can be applied to various facets of their business. The hallmark of this stage is a shift from theoretical interest to practical engagement. Unlike the awareness stage, where people often have abstract or speculative conversations about AI, the exploration stage features targeted research, feasibility studies, and small-scale pilot projects.

At this point, organizations usually have a designated individual or a small team responsible for AI initiatives. This team often collaborates with other departments to identify potential use cases for AI within the organization. Budget allocations for AI are more substantial compared to the awareness stage, but they are still conservative, earmarked for specific projects or proof-of-concept endeavors.

The focus during the exploration stage is on actionable insights. Organizations often engage with external consultants or AI solution providers to conduct feasibility studies or to develop prototypes. These activities are not just experimental but are aligned with specific business objectives, whether it's improving customer engagement, optimizing supply chain logistics, or enhancing product quality.

So, what should organizations at the exploration stage be doing? First, they should be formalizing their AI strategy, outlining clear objectives, KPIs, and timelines. Second, they should be investing in skill development, either by training existing staff or by hiring AI specialists. Third, organizations should consider partnerships with AI vendors or academic institutions to accelerate their learning curve. Lastly, a more substantial budget should be allocated to fund these activities, recognizing that investment at this stage can yield significant long-term benefits.

IMPLEMENTATION

By the time an organization reaches the implementation stage, it has already laid a solid foundation through awareness and exploration. Now, the focus shifts from "Can we do this?" to "How do we scale this?" The defining characteristic of this stage is the rollout of AI solutions that are aligned with the organization's strategic objectives. Unlike the exploration stage, where pilot projects and feasibility studies are the norm, the implementation stage is about deploying AI solutions at scale.

In terms of organizational structure, there's usually a dedicated AI division by this point, often led by a chief AI officer or an equivalent role. This division is not working in isolation but is deeply integrated with other departments, ensuring AI initiatives are aligned with broader business goals. Budgets for AI are not just substantial but are also long-term, reflecting the organization's commitment to AI as a strategic asset.

The key activities at this stage include the scaling of successful pilot projects, continuous monitoring and optimization, and the institutionalization of AI best practices. Organizations often employ advanced data analytics tools to measure the performance of AI implementations against predefined KPIs. Any gaps or inefficiencies are promptly addressed, often leading to iterative cycles of improvement.

So, what should organizations at the implementation stage focus on? First, they need to establish robust governance structures to oversee AI initiatives, ensuring ethical and responsible AI usage. Second, focus should be on talent retention and development, as the demand for skilled AI professionals is high. Third, organizations should be looking at AI not just as a tool for operational efficiency but also as a driver for innovation and competitive advantage.

Reflect on Starbucks and its acclaimed rewards program. The coffee company, through a series of trials, collected extensive data covering aspects like customer contact details, buying patterns, the use of offers, and even mobile device IDs. Over ten years of refining their methods, Starbucks evolved from distributing thirty manually created emails weekly to using AI for dispatching four hundred thousand highly tailored emails instantly.[1]

OPTIMIZATION

At the optimization stage, an organization has successfully integrated AI into its core operations and is now focused on fine-tuning and expanding its AI capabilities. The organization is not just solving existing problems more efficiently but is also identifying new opportunities and avenues for growth that were previously unimaginable. The AI strategy is mature, dynamic, and aligned with the organization's long-term vision.

In terms of organizational structure, the AI division is usually a strategic unit with significant influence over business decisions. The budget for AI is not just large but is also flexible, allowing for rapid adjustments in response to emerging opportunities or challenges. The AI team is highly specialized, often including not just data scientists but also experts in ethics, compliance, and domain-specific knowledge.

The key activities at this stage go beyond mere implementation. Organizations are involved in continuous improvement cycles, leveraging AI to optimize every facet of their business—from customer engagement and employee productivity to supply chain efficiency and innovation. Advanced analytics and real-time dashboards are commonly used to monitor a wide array of performance metrics, enabling data-driven decision-making at an unprecedented scale.

So, what should organizations at the optimization stage focus on? First, they should be looking at scalability across their companies, taking successful AI initiatives and expanding them to different markets or sectors. Second, focus should be on sustainability, ensuring AI initiatives are ethical, compliant, and contribute to broader societal goals. Third, organizations should invest in cutting-edge research and development, staying ahead of the curve in AI technologies and methodologies.

Thread, acquired by the UK-based retailer M&S, is a prime example of a company that has reached the optimization stage. By harnessing AI, the fashion brand has successfully personalized customer wardrobes. This innovation merges insights from personal stylists with data-driven algorithms that analyze a variety of sources. Such a strategy enables Thread to offer customized services to a vast number of clients, significantly surpassing the capabilities of human stylists alone.[2]

TRANSFORMATION

In the transformation stage, an organization has not only optimized its internal operations through AI but has also started to redefine its industry landscape. The focus here is on leveraging AI to create new business models, disrupt traditional value chains, and contribute to societal well-being. AI is deeply ingrained in the organization's DNA, influencing not just strategic decisions but also its mission and vision.

Organizationally, the AI division is often part of a larger innovation hub that collaborates with external stakeholders, including start-ups, academic institutions, and government bodies. The budget for AI is not just substantial but is also earmarked for groundbreaking projects that aim to set new paradigms. The talent pool is diverse, comprised of not just technologists but also thought leaders, ethicists, and domain experts from various fields.

Key activities at this stage are visionary in nature. Organizations are involved in pioneering research, often publishing whitepapers and contributing to academic journals. They are also active in policy advocacy, working with regulators to shape the future of AI governance. Moreover, there's a strong focus on ethical AI, with organizations taking the lead in setting industry standards for responsible AI usage.

So what should organizations at the transformation stage focus on? First, they should aim to be thought leaders in AI, influencing not just industry trends but also public opinion and policy. Second, they should look at collaborative innovation, partnering with external organizations to solve complex, global challenges. Third, they should focus on legacy, considering how their AI initiatives will impact future generations.

DBS Bank, under the leadership of Piyush Gupta, represents a compelling example of a company that has reached the transformation stage of AI maturity. At this level, businesses strategically and pervasively integrate AI across all functions, fundamentally altering the organization's operations and value delivery. Gupta has been instrumental in embracing AI, setting a strong example through his direct involvement in significant data and technology transformations. This level of leadership commitment is crucial in the transformation stage, ensuring AI aligns with the organization's strategic goals and receives the necessary support from top management. He also fostered a culture of innovation and experimentation by establishing KPIs focused on conducting thousands of AI-related experiments annually and encouraging broad employee engagement in innovative practices such as the AWS DeepRacer League.[3]

DBS went through major operational changes to enhance its AI capabilities. The transition from traditional data warehouses to flexible data lakes along with the setup of new governance structures like the Responsible Data Use Committee illustrate DBS's deep integration of AI into its operational framework. Furthermore, the bank has trained over eighteen thousand employees in data skills, promoting a widespread culture of "citizen data scientists," which is vital for supporting and sustaining AI-driven operations. On the strategic front, the bank has effectively deployed AI in various core business areas, such as customer service via advanced chatbots and HR through innovative tools like the Job Intelligence Maestro. These strategic AI deployments not only enhance efficiency but also significantly improve service delivery, marking the transformative impact of AI at DBS. Moreover, Gupta's emphasis on developing in-house AI capabilities to rival digital natives underscores a forward-thinking strategy where AI is leveraged not just for operational efficiency but as a crucial competitive differentiator.[4]

Through these initiatives, DBS has transformed from a traditional bank into a leading, AI-driven financial institution. This transformation exemplifies the advanced stage of AI maturity, where AI is not merely an adjunct technology but a central pillar driving the company's strategic innovation and competitive positioning.

Stage	Description	What Orgs Should Be Doing
Awareness	Basic understanding of AI	Educate employees, assess AI impact, small budget for explorations
Exploration	Targeted research, small pilots	Formalize AI strategy, pilot projects, partnerships, allocate budget
Implementation	Ready to scale AI solutions	Deploy solutions, establish AI division, governance structures, talent development
Optimization	AI integrated into core operations	Global scalability, cutting-edge R&D, continuous improvement, advanced analytics
Transformation	Redefining industry with AI	Thought leadership, collaborative innovation, policy advocacy, pioneering research

Table 9: Five Stages of AI Maturity

ASSESSING YOUR ORGANIZATION'S AI MATURITY

Understanding your organization's AI maturity is crucial for effective planning and development. This section discusses a comprehensive framework to evaluate your current capabilities, identify gaps, and benchmark against industry standards.

FRAMEWORK FOR ASSESSING AI MATURITY

Understanding where your organization stands on the AI maturity curve is crucial for planning your AI journey effectively. It helps you identify gaps, allocate resources wisely, and set realistic goals. Here's a simple yet comprehensive framework to assess your organization's AI maturity.

STEP 1: CONDUCT AN INTERNAL AUDIT

First, you'll want to know what various teams are already doing with AI. Conduct an internal audit to gather a baseline understanding of your organization's current AI capabilities. Here are the steps:

- **Create an inventory:** List all ongoing and completed AI initiatives, their objectives, and outcomes.
- **Assess resources:** Determine the human and financial resources currently dedicated to AI.
- **Evaluate the technology stack:** Check if your existing technology infrastructure can support AI development and deployment, including workloads and the performance of machine learning and language models. For a refresher on building foundations for security and infrastructure, please review chapters 5 and 7.

You can use internal surveys, interviews with key stakeholders, and financial reports to gather this information.

STEP 2: EVALUATE ORGANIZATIONAL READINESS

Once you have a snapshot of your current state, the next step is to evaluate how ready your organization is to adopt and scale AI. Skill assessment tests, data quality tools, and leadership surveys can be useful tools here.

- **Skill assessment:** Evaluate the AI-related skills within your team. We have discussed key players for success of AI in chapter 8.
- **Data readiness:** Assess the quality and availability of data. Additional details about the criticality of data for AI can be found in chapter 6.
- **Leadership alignment:** Ensure the leadership team is aligned with the AI vision and committed to its implementation.
- **Governance:** Assess what governance and oversight policies and procedures are in place. We discussed developing AI responsibly in chapter 9.

STEP 3: BENCHMARK AGAINST INDUSTRY STANDARDS

Now, it's time to look outward. Compare your organization's AI capabilities with those of your competitors and industry benchmarks. This involves:

- **Competitive analysis:** Understand how your competitors are leveraging AI.
- **Industry benchmarks:** Use industry reports to see where you stand in terms of AI maturity.

Market research reports and competitor case studies can offer valuable insights in both areas.

STEP 4: IDENTIFY GAPS AND OPPORTUNITIES

After internal and external evaluations, it's time to identify the gaps and opportunities. Strengths, weaknesses, opportunities, and threats (SWOT) analysis and opportunity assessment frameworks can be particularly useful at this stage.

- **Gap analysis:** Compare your current capabilities with your AI goals to identify gaps.
- **Opportunity mapping:** Pinpoint areas where AI can bring the most value to your organization.

STEP 5: DEVELOP AN ACTION PLAN

Finally, armed with all this information, develop a concrete action plan. Project management software and budgeting tools can help you stay on track.

- **Setting objectives and KPIs:** Clearly define your goals for the next phase of your AI journey.
- **Resource planning:** Allocate the necessary budget and manpower.
- **Timeline:** Establish a realistic timeline for achieving your objectives.

Conduct an Internal Audit

• Gather baseline data on current AI projects, resource allocation, and technology infrastructure

Evaluate Organizational Readiness

• Assess the skills, data readiness, and leadreship aslignment for AI adoption

Benchmark Against Industry Standards

• Compare your AI capabilities with competitors and industry benchmarks

Identify Gaps and Opportunities

• Perform a gap analysis and map out areas where AI can add value

Develop an Action Plan

• Create a concrete plan with objectives, KPIs, resource allocation, and a timeline

Fig 14: Framework for Assessing AI Maturity

THE JOURNEY TOWARD AI MATURITY

The path to AI maturity is a long journey and requires a well-thought-out strategy, a commitment to continuous improvement, and alignment of initiatives with broader business objectives.

- **Strategic Alignment:** One of the first steps in this journey is to ensure strategic alignment. This requires regular consultations between the AI team and top management to make sure the AI projects undertaken serve the larger goals of the organization.
- **Resource Allocation:** Investing in the right talent and technology is paramount for the successful implementation of AI. This could mean hiring AI specialists or providing training programs to upskill your current workforce. Additionally, budget allocation for AI projects should align with your strategic objectives. The aim is to build a robust AI infrastructure that can support your long-term goals.
- **Pilot Projects:** Before fully integrating AI across the company operations, it's wise to conduct pilot projects. These projects serve as a litmus test for the effectiveness and ROI of your AI solutions. They offer valuable insights that can help you refine your approach, ensuring you're on the right path. Pilot testing minimizes the risk of large-scale failures and allows for adjustments to be made before a full-scale rollout.
- **Monitoring and Optimization:** Once your AI solutions are deployed, it's crucial to continuously monitor their performance using advanced analytics tools. These tools can provide real-time insights into various performance metrics, enabling you to make data-driven decisions

for optimization. This cycle of constant improvement is aimed at ensuring your AI initiatives deliver sustained value.
- **Governance and Ethics:** As you scale your AI initiatives, governance and ethics become increasingly important. A governance framework, which sets guidelines for data usage, algorithmic fairness and accountability, ensures your AI solutions are ethical and responsible.

The AI maturity model provides a strategic roadmap for the integration of artificial intelligence into all facets of your organization. While the journey may seem daunting initially, taking it one stage at a time in a structured manner is key. Focus on the specific goals and best practices relevant to your current level of maturity. Lay the groundwork in a robust manner so you can ascend to the next stage seamlessly.

Leverage the collective insights and experiences of organizations that have traversed this path before you. Their successes and setbacks hold valuable lessons. With a sound AI strategy, the right resources and talent, and a commitment to continuous improvement, your organization too can unlock the transformational benefits of AI. Progress through the stages of AI maturity, and you will be well-positioned to disrupt and lead in the age of artificial intelligence.

THE TRANSFORMATIONAL BENEFITS OF AI MATURITY

The benefits of progressing through the stages of AI maturity are manifold. Advanced AI capabilities can provide a significant competitive edge, enabling you to offer superior products or services. Operational excellence is another key benefit, as AI can streamline various business processes—from supply chain management to customer service. This leads to improved efficiency and cost reductions.

Moreover, AI opens the door to innovation, offering new avenues for business growth that may not have been previously considered. On the customer front, AI can significantly enhance the customer experience by personalizing customer journeys, leading to higher satisfaction and loyalty. And let's not forget the scalability advanced AI solutions offer, allowing you to expand your reach across different markets and sectors. At the highest level of AI maturity, you also have the opportunity to make a positive societal impact through ethical and responsible AI usage.

CHAPTER 19

The Future of Work: AI and Human Collaboration

The morning sun filtered through the towering fir trees, casting a warm glow over the blueberry fields outside Portland, Oregon. As I plucked the ripe berries, the tranquil silence was shattered by my phone buzzing—a call from Michelle, my good friend and senior partner at a prestigious executive search firm. Her voice typically exuded self-assured confidence honed over decades of matching top talent with Fortune 500 clients.

"Reddy, be honest with me—is AI going to make professionals like us obsolete?" Michelle's words channeled the fears trickling across industries amid AI's relentless ascent. Could this technological disruption render hard-earned expertise redundant? I measured my response carefully.

"Michelle, I don't subscribe to the narrative that AI will make your role obsolete," I began calmly. "If anything, I see it as a force-multiplier that can elevate your work to new strategic heights." I described AI's potential as a highly capable assistant that could rapidly synthesize nuanced job specs and candidate profiles far beyond any individual's bandwidth. But critically, AI's analytical power would be harnessed to Michelle's uniquely seasoned artistic direction.

"You will be leveraging the data foundation built by AI, which understands the complexities of your field. You can then apply your robust intuition to shape that input into a beautifully tailored solution for your clients."

As I outlined that symbiotic vision of human-AI collaboration, Michelle's apprehension dissipated. The allure of using AI to offload tedious tasks while amplifying her team's strategic value through elevated discernment rekindled her trademark self-assuredness.

"You're right, Reddy," she said with a knowing nod. "AI won't obsolete us. It will be the complementary force propelling our profession to unprecedented new heights."

The future of work sparked fears that AI could make humans obsolete, but we were likely headed toward an era of human-AI collaboration rather than wholesale replacement of workers. While AI would bring tectonic workplace shifts and make some roles redundant, the powerful combination of human ingenuity and artificial intelligence would elevate most professions to new heights.

AI AUTOMATION OF CERTAIN JOBS AND TASKS

While AI will complement humans in many roles, it is also true AI will automate some functions previously handled by humans. Jobs most susceptible to AI automation involve predictable physical activities or data collection and processing. These include positions like retail cashiers, manufacturing assembly line workers, call center operators, and financial analysts.

For example, Amazon Go stores rely on computer vision and sensors to track what items customers pick up, eliminating the need for cashiers.[1] In factories, robotic arms can repeat precisely programmed motions to assemble products.[2] Chatbots are handling more routine customer service inquiries, reducing the need for call center staff.[3] In finance, AI can automate tasks like cash flow analysis, invoicing, financial statement generation and credit risk modeling.[4]

This does not mean these jobs will disappear entirely. But the day-to-day work will change significantly. Humans will have less need to perform repetitive, routine tasks that machines can handle equally or better. Second, there will be AI-driven productivity gains and therefore, the number of people in each function like finance or marketing will be fewer.

ADAPTING TO THE AGE OF INTELLIGENT MACHINES: A GUIDE FOR LEADERS

Leaders will need to take proactive steps to prepare themselves and their workforce for this transformation. Walmart, recognizing many cashier roles will become obsolete, is already reskilling employees to work alongside the autonomous checkout technology.[5] Other companies will need similar training programs to transition staff into more value-add roles.

AI is automating more tasks every day and influencing the way we work. Leaders need to take proactive steps to prepare their workforces for these changes. It is a daunting task, but some key strategies have proven to be effective.

First off, companies need to conduct skills gap analyses to identify roles and tasks likely to be automated. With this understanding, they could develop reskilling and upskilling programs to help employees transition to new roles. These training initiatives should focus on soft skills like creativity, empathy, and collaboration—qualities that would become increasingly valuable as machines took over more routine tasks.

Second, companies need to foster a culture of continuous learning, instilling adaptability and willingness to learn new skills as workforce needs evolved. Also, recruiting and talent strategies need to be updated, shifting the focus toward hiring for potential rather than specific technical skills that could quickly become obsolete.

Third, to expose employees to different functions and skills, companies should implement job rotation programs. Additionally, they should provide opportunities for employees to gain hands-on experience with AI tools through pilots and experiments, enabling smoother integration.

Lastly, it's essential to maintain open communication about how AI might impact jobs. This can help alleviate uncertainty and build trust within the workforce. If permanent roles are reduced, leaders should consider alternative work models such as partnerships and contracting.

Ultimately, the key is to focus on augmenting human strengths of creativity, judgment, and reasoning. By developing complementary roles that leverage the unique abilities of both humans and machines, companies can make the most of AI's automation capabilities while ensuring their employees remain valuable and engaged.

With the right preparation and training, the narrative would unfold where companies and their workforces navigate the changes brought about by AI automation, emerging stronger and more adaptable on the other side.

THE CONTINUED NEED FOR HUMAN ABILITIES

While certain routine and repetitive tasks will be automated, AI still lacks core human abilities that will remain critical in the workplace. As MIT professors Erik Brynjolfsson and Tom Mitchell noted, "It's become increasingly clear that human skills, often called soft skills, will become increasingly important."[6] These inherently human skills include creativity, empathy, judgment, cross-domain thinking, and the ability to manage interpersonal relationships.

Creativity involves imaginative problem-solving and ideation that machines cannot replicate. As Martin Ford, futurist and author, says, "The areas that, at least for the foreseeable future are going to be less susceptible to automation, are things that are genuinely creative, where you're really thinking outside the box, creating something new."[7] Humans will continue to be needed to pioneer innovative products, services, marketing campaigns, and new business models.

Empathy and emotional intelligence allow humans to understand unspoken needs and connect with others at an emotional level. The easiest jobs to automate are those requiring low amounts of human interaction and emotional intelligence. As companies rely more on data and analytics, they will need employees who can relate to customers and make sound people decisions.

Human judgment and intuition will also remain essential for complex, ambiguous decisions. When faced with imperfect or incomplete data, subjective human assessment and experience are invaluable. For example, DeepMind's AlphaFold can predict protein structures with remarkable accuracy. But human intervention was still needed to select the right protein model when AlphaFold was trained with limited data. Hybrid teams that combine AI and human perspectives make the best judgments. In critical decision-making scenarios such as loan approvals, health care, and hiring, humans are indispensable. By drawing connections across diverse domains, humans provide the cross-domain thinking AI models lack. Seeing patterns Google's AlphaGo missed allowed human Go champion Lee Sedol to occasionally beat the AI system.[8] Humans excel at transferring insights between areas.

Leaders will need to nurture these human strengths as they incorporate more AI capabilities into their organizations. Roles focused on innovation, relationships, and multifaceted thinking will become increasingly important. Further, human oversight remains essential for maintaining transparency and upholding ethical standards, which ensures decisions align with business values and benefit stakeholders.

EMERGING ROLES FOR HUMANS ALONGSIDE AI

As AI becomes more pervasive in the workplace, new roles are emerging that connect the complementary strengths of humans and machines. These roles focus on combining human creativity, empathy, judgment, and strategic thinking with AI's data processing, pattern recognition, and predictive capabilities. Here are several key emerging roles:

AI Strategist: Responsible for aligning AI deployment with overall business strategy, AI strategists bridge the gap between technical and nontechnical stakeholders. They identify areas where AI can deliver value and develop strategic roadmaps that incorporate the technology into broader business objectives. For example, an AI strategist at a retail company might develop a plan to use AI for demand forecasting, inventory optimization, and personalized product recommendations.

Ethics and Governance Specialist: This role involves ensuring AI systems are used ethically and responsibly. Specialists in this area develop guidelines and frameworks to address issues like bias, transparency, data privacy, fairness, and accountability, ensuring AI aligns with organizational values and regulatory standards. They might conduct algorithmic audits to identify and mitigate sources of bias or develop privacy policies for handling sensitive data used to train AI models.

Hybrid Team Coordinator: These professionals manage teams where AI tools and human experts work side by side. They optimize workflows by identifying tasks best handled by humans or machines and cultivate a culture of collaboration to maximize overall team productivity and innovation. In a health-care setting, a hybrid team coordinator might oversee a team where AI assists radiologists in identifying potential abnormalities in medical images while the radiologists provide final diagnoses and recommend treatment plans.

Human-Machine Collaboration Designer: In this role, individuals design user interfaces and workflows that enhance human-AI collaboration. They ensure AI tools are intuitive and accessible, enabling people to make the most of AI's capabilities. For example, they might develop a natural language interface that allows nontechnical users to query an AI system for insights or recommendations.

AI Trainers and Data Curators: AI systems need accurate and relevant data to perform well. Trainers and curators refine datasets, manage data quality, and provide feedback to improve AI models. They also train AI systems to recognize nuances in data, refining them for specific applications. In the field of natural language processing, data curators might annotate text data to help train AI models to better understand context and sentiment.

AI Explainers: AI explainers specialize in translating technical AI concepts into accessible language for nontechnical audiences. They help stakeholders understand AI's decisions and implications, building trust in the technology. For example, an AI explainer might present the rationale behind an AI-driven loan approval decision to a bank customer in clear, understandable terms.

Innovation Catalyst: These professionals inspire and guide cross-functional teams to develop new ideas and innovations, leveraging AI to create groundbreaking products and services. They encourage experimentation and challenge conventional thinking to unlock the full potential of human-AI collaboration. In the automotive industry, an innovation catalyst might lead a team exploring the use of AI for autonomous driving or predictive maintenance.

AI Training Specialist: As AI becomes more prevalent, there is a growing need for professionals who can develop and deliver educational programs on AI for employees, customers, and other stakeholders. These specialists design curricula, create training materials, and facilitate workshops to build AI literacy and help people effectively work with and understand AI systems.

These roles emphasize that the future of work involves seamless human-AI collaboration. As businesses evolve, leaders will need to recognize the significance of these emerging roles and nurture talent that can thrive at this intersection, fostering a culture of continuous learning and adaptation.

PREPARING THE WORKFORCE FOR HUMAN-AI COLLABORATION

To fully leverage the strengths of both humans and AI, companies need to deliberately prepare their workforce for seamless collaboration between man and machine. Here are some best practices for developing the skills needed for human-AI hybrid teams.

Foster Constant Learning: Encourage employees to acquire "fusion skills." As automation changes job roles, workers must continuously gain skills that fuse technical fluency with strengths like creativity and empathy. Managers should nurture cross-functional knowledge and varied experience. Incorporate emotional intelligence training to nurture unique human skills like psychological safety, relationship-building, and communication. Classes on inclusivity and bias mitigation will also enable ethical AI oversight.

Make Room for Creativity: Provide opportunities for creative exploration and exposure to interdisciplinary ideas. Encourage collaboration between departments like HR, IT and operations. Break down data silos so all employees gain comfort analyzing data and understanding AI models. This facilitates teaming across disciplines.

Start Small with AI: Let employees get first-hand experience while limiting risk by piloting AI in one business unit. Learn what works through these experiments before scaling AI across the entire company. For example, Spotify adopted a collaborative, experimental approach as they developed AI systems to personalize music recommendations. Engineers and music experts worked closely together, ensuring the AI met both technical and creative goals. This fusion approach allowed Spotify to create an AI that exceeded either discipline alone.[9]

Transitioning to a future of humans and machines working together requires concrete actions to equip the workforce with the right mindset and skillset. But done properly, integrated human-AI collaboration can uplift workers and organizations alike.

MANAGING HUMAN-AI TEAMS

To realize the full potential of human-AI collaboration, companies must adopt management practices that seamlessly integrate both human and machine capabilities.

BEST PRACTICES FOR MANAGING HYBRID TEAMS

Managing hybrid teams, where AI systems and humans work together, requires careful planning to ensure seamless collaboration. Here are some best practices.

Define Roles and Responsibilities: Clearly defining the roles and responsibilities of both systems and staff is essential. Conduct collaborative workshops to map out workflows, decision rights, and handoff points. This process ensures everyone understands their specific contributions and knows when to step in or hand over tasks to AI systems.

Establish Feedback Loops: Constructive feedback loops between AI systems and end users are crucial for continuous improvement. Building interfaces that enable easy tagging of AI predictions as accurate or inaccurate allows users to provide valuable feedback, retraining models and enhancing their accuracy over time.

Maintain Human Oversight: Human oversight remains paramount in hybrid teams. Implement approval workflows and monitoring dashboards to track AI performance. Empower human staff to override incorrect AI decisions and prevent harmful outcomes by setting appropriate thresholds for manual intervention.

Promote Two-Way Learning: Encourage two-way learning where AI informs humans and vice versa. Regular working sessions between AI developers and business users foster collaboration as they review model outputs together, exchanging insights to refine both AI systems and human understanding.

Foster Diversity: Diverse and multidisciplinary teams are critical for building robust AI systems. Use inclusive hiring practices to ensure developers are exposed to users from various backgrounds. Teams should include domain experts, ethicists, and UI/UX designers who can approach challenges from different angles and build more effective systems.

Properly implemented human-AI collaboration does not diminish human roles but rather elevates both man and machine. Thoughtful management of hybrid teams unlocks tremendous potential.

Collaboration between humans and AI systems, each contributing their unique strengths, will define the future of work. Many new roles for humans are emerging to train, explain, sustain, and design AI systems. Preparing for this future will require changes in how we develop our workforce. Employees will need to continuously gain "fusion skills" that complement AI capabilities. Constant learning, emotional intelligence, creativity, and cross-disciplinary exposure will be critical. With the right mindset and skills, humans have immense opportunities to thrive alongside increasingly intelligent machines.

By smoothly navigating the transition and new ways of working, businesses can enter an era of uplifted productivity, innovation, customer service, and employee satisfaction through integrated human-machine teams.

CHAPTER 20

The Road Ahead

"The illiterate of the twenty-first century will not be those who cannot read and write, but those who cannot learn, unlearn, and relearn." —Alvin Toffler

Artificial intelligence (AI) is dramatically changing the way we live and work, ushering in an era of innovation, disruption, and growth. As we've seen, AI is having a major impact on various industries, empowering organizations to rethink their processes and challenging traditional business models. From manufacturing and logistics to health care and finance, AI's influence is undeniable.

In earlier chapters, we examined how AI-driven automation and predictive analytics are unlocking unprecedented efficiencies. Manufacturing floors are becoming smarter as intelligent robots perform complex tasks with precision, reducing human error and freeing workers to focus on creative and strategic activities. In health care, AI is offering diagnostic tools that improve accuracy and provide personalized treatments, potentially transforming patient care globally.

This revolution extends beyond individual industries to the very fabric of society. Generative models are enabling creative professionals to innovate in art, design, and literature. Customer service is being redefined through conversational agents that learn from each interaction. Meanwhile, predictive models are helping organizations anticipate market trends, optimize supply chains, and adapt to shifting consumer preferences.

Yet with these immense possibilities come challenges that cannot be ignored. The workforce must evolve to remain relevant in this new age, and businesses must grapple with ethical dilemmas, ranging from bias in AI systems to ensuring data privacy. Our journey so far has revealed that the adoption of AI is no longer optional but imperative. Those who fail to harness this potential risk being left behind as the world races toward an AI-powered future.

THE FUTURE OF AI: TRENDS AND PREDICTIONS

The field of artificial intelligence is rapidly evolving, and the next decade promises to bring about remarkable advancements and transformative applications. As we look ahead, several key trends and predictions emerge, shaping the future trajectory of AI.

ADVANCEMENTS IN AI CAPABILITIES

One of the most significant developments on the horizon is the progress toward artificial general intelligence (AGI) or strong AI. While current AI systems excel at specific tasks, AGI aims to create machines with human-level intelligence capable of reasoning, learning, and solving problems across diverse domains. Achieving AGI could lead to breakthroughs in areas such as natural language processing, decision-making, and problem-solving, enabling AI systems to tackle complex, multifaceted challenges.

Another exciting frontier is the rise of multimodal AI, which combines and processes multiple forms of data, including text, images, audio, and video. By integrating these modalities, AI systems can develop a more comprehensive understanding of the world, leading to enhanced capabilities in areas like virtual assistants, robotics, and multimedia analysis.

Natural language processing (NLP) is also poised for significant advancements, with AI systems becoming increasingly adept at understanding and generating human-like language. This will enable more natural interactions with virtual assistants, chatbots, and other language-based applications, leading to significant advancements in customer service, content creation, and language translation.

EXPANSION OF AI APPLICATIONS ACROSS INDUSTRIES

As AI capabilities continue to advance, we can expect to see an unprecedented expansion of applications across virtually every industry. AI-driven robotics will play a crucial role in areas like manufacturing, logistics, and health care with robots becoming more autonomous, collaborative, and capable of handling complex tasks.

AI has the potential to dramatically transform health care by significantly improving diagnostics, drug discovery, and personalized medicine. In finance, AI-powered systems will drive intelligent trading, risk management, and fraud detection. Manufacturing and logistics will benefit from predictive maintenance, optimized supply chains, and autonomous systems.

We will also likely see a pivotal role for AI in addressing global challenges such as climate change, energy efficiency, and sustainable development. By analyzing vast amounts of data and simulating complex scenarios, AI can help identify solutions for reducing emissions, optimizing resource utilization, and mitigating the impact of climate change.

The integration of AI with blockchain technology will lead to new applications in areas like supply chain management, digital identity, and decentralized finance. AI can help analyze and secure blockchain transactions while blockchain can provide a tamper-proof and transparent environment for AI systems.

AI will elevate immersive technologies like virtual reality (VR), augmented reality (AR), and mixed reality (MR), making experiences more realistic and interactive. AI is poised to dramatically reshape gaming, education, and training by introducing incredibly realistic simulations, intelligent virtual assistants, and adaptable learning environments.

QUANTUM COMPUTING AND ITS IMPACT ON AI

Quantum computing, a landmark technology that leverages the principles of quantum mechanics, has the potential to supercharge AI capabilities. By leveraging quantum algorithms and processing power, AI systems could tackle computational problems that are intractable for classical computers, leading to breakthroughs in areas such as cryptography, optimization, and simulation.

The integration of quantum computing and AI could unlock new frontiers in fields like drug discovery, materials science, and financial modeling, where complex calculations and simulations are essential. However, realizing the full potential of quantum AI will require significant advancements in hardware, software, and algorithm development.

As we look to the future, the possibilities presented by AI are both exhilarating and humbling. While the path ahead is filled with challenges and uncertainties, the relentless pace of innovation and the collective efforts of researchers, developers, and industry leaders will undoubtedly forge a future where AI plays an increasingly pivotal role in reimagining our world.

NEW BUSINESS MODELS: AI-DRIVEN INNOVATION

Artificial intelligence is ushering in new business models that redefine how organizations create, deliver, and capture value. Here are some notable trends:

DATA MARKETPLACES

AI depends on vast amounts of quality data to train and refine its models. This has led to the rise of data marketplaces, where organizations can buy and sell datasets to enhance their algorithms. Companies that gather and curate unique data are now tapping into new revenue streams by offering anonymized, aggregated insights to other organizations.

For instance, in the health-care sector, aggregated, anonymized health-care data is helping research institutions and pharmaceutical companies accelerate drug discovery and better understand disease patterns.[1] In the marketing world, consumer behavior data is often purchased to refine customer segmentation and improve targeted advertising campaigns.[2]

INTELLIGENT PRODUCT RECOMMENDATIONS

E-commerce platforms use AI to deliver personalized product recommendations, thereby boosting sales and customer loyalty. By analyzing customers' past purchases, browsing history, and stated preferences, these platforms can suggest complementary products and curate collections that enhance the customer journey.[3] Retailers can increase cross-selling and upselling opportunities by offering customers highly relevant recommendations.[4]

AUTONOMOUS SYSTEMS

Autonomous systems capable of learning and adapting without human intervention are transforming industries like transportation and logistics. Self-driving vehicles and delivery drones are changing traditional logistics models, making delivery services more efficient and flexible.

Logistics companies are investing heavily in autonomous vehicles and drones to ensure goods are delivered quickly and cost-effectively in urban areas.[5] In manufacturing, factories are incorporating AI into production lines that can self-optimize, minimizing waste and maximizing efficiency.[6]

PLATFORM ECOSYSTEMS

AI is fueling the growth of platform ecosystems that connect buyers and sellers in new ways. Predictive analytics helps these platforms better match supply with demand, reducing transaction friction and creating a more seamless customer experience.

Gig economy platforms like Uber and TaskRabbit utilize AI to match workers to jobs efficiently, ensuring services are available when needed. Business-to-business marketplaces also harness machine learning to recommend suppliers based on purchasing patterns and market trends.[7]

PEER-TO-PEER MODELS

AI is also catalyzing the growth of peer-to-peer (P2P) business models, where individuals or small groups can directly exchange goods and services. These P2P models are upending traditional intermediaries, fostering direct buyer-seller interactions predicated on transparent, data-driven reputational metrics that cultivate trust.

For instance, peer-to-peer lending platforms use AI to assess borrowers' creditworthiness more accurately, enabling individuals to access funding directly from other people rather than traditional financial institutions. Similarly, peer-to-peer marketplaces allow users to share and exchange resources like homes, cars, and even personal skills, often facilitated by AI-enhanced algorithms that provide secure, fair transactions.[8]

PREPARING FOR THE AI-POWERED FUTURE

As AI technologies accelerate, it is imperative for businesses, governments, and societies to proactively prepare for the transformative changes that lie ahead. Embracing the AI-powered future requires a multifaceted approach, encompassing workforce development, cultural shifts, and strategic investments.

BUILDING AN AI-READY WORKFORCE

One of the most critical components of preparing for the AI-powered future is cultivating a workforce with the necessary skills and mindset to thrive in an AI-driven environment. This endeavor involves investing in AI education and training programs to equip workers with the knowledge and technical expertise required to develop, deploy, and maintain AI systems. It also necessitates fostering cross-functional collaboration by bringing together professionals from diverse fields, including computer science, data science, domain expertise, and ethics.

FOSTERING A CULTURE OF CONTINUOUS LEARNING AND ADAPTABILITY

Beyond functional skills, organizations must cultivate a culture that embraces continuous learning and adaptability. This involves encouraging a growth mindset and a willingness to experiment, fail, and learn from mistakes in the pursuit of AI innovation. Promoting cross-functional collaboration and knowledge-sharing is essential to facilitate the dissemination of AI expertise throughout the organization.

INVESTING IN AI EDUCATION AND RESKILLING INITIATIVES

Governments, educational institutions, and private organizations must collaborate to establish comprehensive AI education and reskilling initiatives. These efforts should focus on updating educational curricula at all levels—from primary and secondary schools to universities—to incorporate AI concepts, ethical considerations, and practical applications. Developing vocational training programs and bootcamps will equip workers with AI-related skills in areas such as data science, machine learning, and AI engineering. Encouraging public-private partnerships will bridge the gap between academia and industry, ensuring that educational programs align with the evolving needs of the workforce. Providing accessible and affordable reskilling opportunities for workers displaced by automation and AI will enable them to transition to new roles and industries seamlessly.

ETHICAL CONSIDERATIONS AND GOVERNANCE

As AI systems become increasingly sophisticated and integrated into various aspects of our lives, ethical considerations and robust governance frameworks are vital to ensure their responsible development and deployment. The rapid advancement of AI technology has outpaced the formulation of comprehensive guidelines and regulations, underscoring the urgency to address these critical issues head-on.

ADDRESSING BIAS AND FAIRNESS IN AI SYSTEMS

One of the most pressing ethical concerns surrounding AI is the potential for perpetuating and amplifying societal biases, leading to discriminatory outcomes. AI systems inherit and exacerbate biases present in the data used to train them, resulting in unfair and harmful consequences in areas such as employment, lending, and criminal justice systems.

Addressing bias requires a multifaceted approach that involves developing techniques to detect and mitigate bias in training data and AI models. It also necessitates promoting diversity and inclusivity within the teams developing AI systems to bring diverse perspectives and experiences to the table. Establishing robust frameworks and metrics to assess the fairness and nondiscrimination of AI systems is crucial. Furthermore, encouraging transparency and explainability in AI decision-making processes will enable scrutiny and accountability, fostering trust in these technologies.

ENSURING TRANSPARENCY AND ACCOUNTABILITY

Many AI systems are opaque and often referred to as "black boxes." Without a clear understanding of how these systems arrive at their decisions, it becomes increasingly challenging to assign responsibility and ensure proper oversight. Addressing this issue involves developing explainable AI models that can provide insights into their decision-making processes, enabling meaningful human oversight. Establishing clear lines of responsibility and accountability for the development, deployment, and outcomes of AI systems is essential. Implementing robust auditing and monitoring mechanisms will help detect and mitigate unintended consequences or violations.

REGULATORY FRAMEWORKS FOR RESPONSIBLE AI DEVELOPMENT

As AI continues to pervade various aspects of society, we face a growing need for comprehensive regulatory frameworks and guidelines to govern its development and use. These efforts should focus on developing robust AI governance frameworks that strike a balance between fostering innovation and addressing ethical considerations while prioritizing societal well-being. Establishing industry-specific guidelines and best practices for AI development and deployment in sectors such as health care, finance, and transportation is crucial. Encouraging international collaboration and harmonization of AI regulations will ensure consistent standards and facilitate cross-border cooperation. Involving diverse stakeholders—including policymakers, industry leaders, ethicists, and the public—in the development of AI governance frameworks is vital to ensure their effectiveness and broad acceptance.

AI AND SUSTAINABILITY

As the world grapples with the pressing challenges of climate change, resource depletion, and environmental degradation, artificial intelligence presents a powerful tool to drive sustainability efforts forward. By leveraging the immense capabilities of AI systems to process vast amounts of data, identify patterns, and optimize complex systems, we can unlock new solutions that promote sustainable development and environmental conservation.

ADDRESSING CLIMATE CHANGE

AI can play a pivotal role in tackling climate change and environmental challenges through a variety of applications. First, AI algorithms can analyze vast datasets from Earth observation satellites, weather stations, and climate models to improve our understanding of climate patterns, predict extreme weather events, and inform mitigation and adaptation strategies. Second, AI can optimize energy systems by predicting energy demand, managing smart grids, and improving the efficiency of renewable energy sources such as wind and solar power. Third, AI-powered precision agriculture techniques can optimize crop yields, reduce water and pesticide usage, and promote sustainable farming practices through data-driven decision-making. Fourth, AI-enabled remote sensing and image recognition can monitor deforestation, track wildlife populations, and identify areas in need of conservation efforts. Finally, AI systems can optimize waste collection routes, sort and classify waste materials, and identify opportunities for resource recovery and circular economy models.

DEVELOPING ENERGY-EFFICIENT AI SYSTEMS

While AI holds immense potential for driving sustainability, the energy consumption and carbon footprint of AI systems themselves must be addressed. Training large AI models can consume vast amounts of computational power, contributing to greenhouse gas emissions and environmental impact. To mitigate this, efforts should focus on developing energy-efficient hardware and software architectures for AI systems as well as leveraging techniques such as pruning, quantization, and model compression. Encouraging the use of renewable energy sources to power AI data centers and computational infrastructure is crucial. Establishing industry standards and best practices for measuring and reporting the environmental impact of AI systems throughout their life cycle is also essential. Furthermore, investing in research and development of novel AI algorithms and architectures that are inherently more energy-efficient and environmentally friendly will be a key step forward.

THE FUTURE OF HUMAN-AI COLLABORATION

The rise of artificial intelligence is reshaping the workforce and redefining the nature of work itself. As automation increasingly handles routine tasks, new job roles and skill requirements will emerge, necessitating a comprehensive reimagining of workforce development strategies. Businesses and policymakers must proactively plan for this transition, ensuring workers are equipped with the skills needed to thrive in an AI-augmented environment.

Instead of viewing AI as a replacement for human labor, the future lies in effective collaboration between humans and AI systems. AI can augment and enhance human capabilities by automating routine tasks, processing vast datasets, and providing intelligent insights. This synergy unlocks new levels of productivity, creativity, and problem-solving abilities, empowering humans to focus on higher level cognitive tasks and strategic decision-making.

For successful human-AI collaboration, building trust in AI systems is paramount. Transparency in AI decision-making processes, clear lines of accountability, and open communication between humans and AI systems are crucial. Moreover, developing intuitive and user-friendly interfaces will promote seamless collaboration and foster widespread adoption.

As AI becomes more capable and autonomous, maintaining human oversight remains imperative. This requires humans to stay in the loop by reviewing AI recommendations, interpreting outputs, and validating predictions. For instance, legal research tools like Luminance have humans verify documents flagged by algorithms, ensuring accuracy and accountability. Additionally, human oversight means designing systems that allow humans to override AI when necessary. An autonomous vehicle encountering a scenario its algorithms cannot accurately judge should shift control smoothly to a human driver, prioritizing safety over efficiency.

Ultimately, oversight requires a mindset that views AI as amplifying human intelligence rather than replacing it. Chess grandmaster Garry Kasparov illustrated this concept of hybrid intelligence with freestyle chess, where humans and AI collaborate to leverage their respective strengths.[9] The notion of AI augmenting humans should be woven into system design.

Effective oversight stems from a culture of transparency and accountability. Transparency around AI use cases and providing feedback avenues foster trust. Humans remain the moral compass guiding technology's course, and through active, ethical oversight, we can ensure AI is harnessed for the benefit of all.

A CALL TO ACTION

The rapid rise of AI offers a unique chance for organizations to propel growth, achieve significant impact, and unlock entirely new levels of efficiency, innovation, and competitive edge. However, responsible and ethical AI development practices are imperative. Addressing algorithmic bias, data privacy, transparency, and accountability will build trust and ensure equitable distribution of AI's benefits across society. Failure to uphold these principles risks unintended negative consequences and erodes public confidence.

This AI revolution requires collaboration and knowledge-sharing among diverse stakeholders: researchers, developers, policymakers, industry leaders and more. Fostering open dialogue, sharing best practices, promoting cross-disciplinary collaboration, and encouraging global cooperation will accelerate AI's positive impact on humanity and society.

THE ODYSSEY CONTINUES: AN AI-ENABLED FUTURE

As explored throughout this book, AI has immense transformative potential across business facets—from predictive analytics to hyper-personalization. Leading companies are already employing AI for data-driven decisions, predictive insights, real-time strategy adjustments, and tailored products and services at unprecedented scales.

However, successfully leveraging AI requires diligent planning, robust data practices, security focus, responsible and inclusive development, and progressing through AI maturity stages to build strategic vision, talent, and infrastructure.

The AI journey has just begun. To stay competitive, businesses must continually evolve by embracing innovations, fostering a culture of learning and experimentation, and contributing to societal good.

While AI represents the "technology of tomorrow," its current capabilities pale in comparison to its potential decades ahead. Just as the computers from the early 1940s seem primitive by today's standards, current advancements in AI are merely small strides in humanity's timeless progress.

To forge ahead, we must persist with the trailblazing spirit and fearless imagination of AI pioneers who transformed "thinking machines" from dreams to reality. We need unrelenting inquiry, invention, openness to interim failures, cross-border collaboration, and courage to traverse the uncharted.

Our embrace of AI has just begun reshaping society. We will define AI's future by the values we instill in it. Guided by hope, wisdom, and humanity, we will make the greatest AI discoveries, continually transforming ourselves along this profound journey.

Glossary

Activation Function: A mathematical function in a neural network that determines the output of a neuron based on its input. Common activation functions include the sigmoid, tanh, and ReLU.

Algorithm: A set of rules or instructions a computer follows to perform a task. In the context of AI, algorithms are used to find patterns or regularities in data.

Artificial Neural Network (ANN): A computing system inspired by the neural networks found in the human brain. It consists of interconnected nodes (analogous to neurons) and is used for tasks like image and speech recognition.

Attention Mechanism: In deep learning, especially in the context of sequence-to-sequence models, this method allows the model to focus on certain parts of the input rather than using the entire fixed-size context at each step.

AutoML: Automated machine learning, commonly known as AutoML, is the process of automating the time-consuming, iterative tasks of machine learning model development. It encompasses the entire pipeline from the raw dataset to the deployment of a machine learning model. AutoML allows data scientists, analysts, and developers to build ML models with high scale, efficiency, and productivity while maintaining model quality.

Backpropagation: A method used in training neural networks, where the algorithm adjusts the weights based on the error of the output.

Bias: In machine learning, bias can refer to two things—the intercept term in linear models and the error introduced by approximating a real-world problem, which may lead to underfitting.

Chatbot: A software application designed to simulate human conversation. It can interact with a user in a natural language like English.

Convolutional Neural Network (CNN): A class of deep neural networks, primarily used in image and video recognition tasks. They process data with a grid-like topology, such as an image.

Cost Function: A function that measures the difference between the predicted output and the actual output. The goal in training a model is to minimize this function.

Data Augmentation: A technique used to increase the amount of training data by slightly altering the original dataset. Common methods include rotations, flips, and zooms on images.

Data Fabric: An architecture that facilitates the integration of various data pipelines and cloud environments using intelligent and automated systems. It aims to unify disparate data systems, embed governance, strengthen security and privacy measures, and provide more data accessibility.

Deep Learning: A subset of machine learning that uses neural networks with many layers (deep neural networks) to analyze various factors of data.

Dropout: A regularization technique for neural networks. During training, randomly selected neurons are ignored, which helps in preventing overfitting.

Embedding Layer: A layer in neural networks that turns positive integers (indexes) into dense vectors of fixed size, often used to process words.

Ensemble Learning: A machine learning paradigm where multiple models are trained to solve the same problem and then are combined to get better results. Examples include Random Forest and gradient boosting machines.

Epoch: One complete forward and backward pass of all the training examples in a neural network.

Feature Engineering: The process of selecting and transforming variables when creating a predictive model. It's an art as much as a science.

Feature Extraction: The process of transforming raw data into a set of inputs (features) that can be easily understood and analyzed by algorithms.

Feedforward Neural Network: A type of neural network where the connections between the nodes do not form a cycle.

Gated Recurrent Unit (GRU): A type of recurrent neural network that has gating mechanisms similar to LSTM but with fewer parameters.

Genetic Algorithm: A search heuristic inspired by the process of natural selection. It's used to find approximate solutions to optimization and search problems.

Gradient Descent: An optimization algorithm used to minimize the cost function by adjusting the model's weights.

Heuristic: A technique designed for solving a problem more quickly when traditional methods are too slow.

Hyperparameter: Parameters whose values are set before training a machine learning model. Examples include learning rate, batch size, and number of epochs.

Image Recognition: The process by which software identifies and categorizes objects within an image.

Imputation: The process of replacing missing data with substituted values, often the mean, median, or mode.

Inference: The process of making predictions using a trained machine learning model.

Jitter: Random noise or disturbance added to data to prevent overfitting and improve generalization.

Kernel: In the context of machine learning, especially support vector machines, a kernel is a function used to compute the dot product in a transformed feature space.

Knowledge Graph: A knowledge base used by Google to enhance its search engine's results with information gathered from various sources.

Long Short-Term Memory (LSTM): A type of recurrent neural network architecture that is well-suited for learning from long-term sequences.

Loss Function: A measure of how well a machine learning model is performing. It quantifies how far off the predictions are from the actual values.

Machine Learning (ML): A type of AI that enables a system to learn from data without being explicitly programmed.

Meta-learning: The idea that we can design models that learn how to learn rather than just learning from data.

Mixture of Experts Model (MoE): A machine learning technique where multiple expert networks are used to divide a problem space into homogeneous regions. It involves decomposing predictive modeling tasks into subtasks, training an expert model on each, and developing a gating model that learns which expert to trust based on the input to be predicted before combining the predictions.

Model Generalization: The ability of a machine learning model to perform well on new, unseen data.

Natural Language Processing (NLP): A field of AI that focuses on the interaction between computers and humans through natural language.

Outlier: A data point that differs significantly from other observations, possibly due to variability in the data or errors.

Overfitting: A modeling error in statistics where a function corresponds too closely to a particular dataset and may fail to fit additional data.

Precision and Recall: Precision is the number of correct positive results divided by the number of all positive results. Recall is the number of correct positive results divided by the number of positive results that should have been returned.

Reinforcement Learning: A type of machine learning where agents learn how to behave by performing actions and receiving rewards.

Regularization: A technique used to prevent overfitting by adding a penalty to the loss function.

Semantic search: A search engine technology that interprets the meaning of words and phrases. It seeks to improve search accuracy by understanding the searcher's intent and the contextual meaning of terms as they appear in the searchable dataspace, whether on the web or within a closed system, to generate more relevant results.

Sequence-to-Sequence Model: A type of model used in tasks like machine translation where both input and output can be of variable length.

Supervised Learning: A type of machine learning where the algorithm is trained on a labeled dataset, meaning the algorithm is provided with input-output pairs.

Tensor: A mathematical object generalized from scalars, vectors, and matrices. In deep learning, tensors are used as the primary data structure.

Transfer Learning: A machine learning method where a model developed for a task is reused as the starting point for a model on a second task.

Underfitting: A modeling error in statistics where a machine learning model is too simple to capture the underlying structure of the data.

Unsupervised Learning: A type of machine learning where the algorithm is trained on an unlabeled dataset, meaning the algorithm explores the data on its own to find patterns.

Vector Database: A vector database is a type of database designed to handle vector data, multidimensional points generated by AI models. It enables efficient storage and retrieval of these vectors for fast and accurate search of similar assets. This makes vector databases crucial in powering visual, semantic, and multimodal searches and use cases.

Virtual Assistant: A software agent that can perform tasks or services based on verbal commands.

Weight: In neural networks, a weight represents the strength of the connection between two nodes.

Word Embedding: A type of word representation that captures the semantic meaning of words based on their context in sentences. Examples include Word2Vec and GloVe.

Xavier Initialization: A method to initialize the weights of a neural network to ensure that the variance remains the same across all layers.

Zero-shot Learning: A type of machine learning where the model correctly makes inferences on data classes it hasn't seen during training.

Z-normalization: A statistical method for normalizing data. It involves subtracting the mean and dividing by the standard deviation.

WHAT ARE TOKENS ANYWAY?

In LLM models, tokens are individual units of text that are used to represent words, phrases, or even whole sentences. When these tokens are converted back into words, the process is called tokenization. There are a few different ways to tokenize text, but the most common approach is to use a word-level tokenizer. This type of tokenizer simply splits the text into individual words, without regard for punctuation or other symbols.

For example, the sentence "The quick brown fox jumps over the lazy dog" would be tokenized into the following words:

The quick brown fox jumps over the lazy dog

Another common approach to tokenization is to use a subword-level tokenizer. This type of tokenizer divides the text into smaller units, such as syllables or morphemes. This can be helpful for languages with complex morphology, such as German or Russian.

For example, the sentence "The quick brown fox jumps over the lazy dog" would be tokenized into the following subwords:

The/qu/ick/bro/wn/fox/jump/s/over/the/lazy/dog

Finally, it is also possible to use a character-level tokenizer. This type of tokenizer divides the text into individual characters. This can be helpful for languages with a large number of characters, such as Chinese or Japanese.

For example, the sentence "The quick brown fox jumps over the lazy dog" would be tokenized into the following characters:

T h e q u i c k b r o w n f o x j u m p s o v e r t h e l a z y d o g

The choice of which type of tokenizer to use depends on the specific application. For example, word-level tokenization is typically used for natural language understanding tasks while subword-level tokenization is often used for natural language generation tasks.

Acknowledgments

I am filled with immense gratitude for the tremendous support that made this book a reality. This book would not have been possible without the invaluable contributions of so many remarkable individuals.

First and foremost, I owe a profound debt of gratitude to my family. My heartfelt thanks go to my wife for her immense patience and encouragement, enduring long days of my book journey. I am eternally grateful to my two children for their immense faith and staunch support throughout this journey.

I am also deeply indebted to my mentors—Scott Herren, Mark Templeton, Bill Siu, and Ken Robinson—whose humility and exceptional leadership have profoundly influenced me. Their encouragement has been a driving force for me to write this book, share my knowledge, and pay it forward.

For this book, I had the privilege of interviewing several esteemed leaders whose profound insights, wisdom, and experiences greatly enriched its content: Joel Hyatt, Deepak Dube, Mandy Long, Trevor Rodrigues-Templar, Maynard Webb, Christian Palmaz, Mike Haske, John Michelson,

Gokula Mishra, George Alifragis, Artem Pochechuev, Wendy Chin, Prof. Angelo Camillo, Manish Gupta, Margot Carlson Delogne, and Farhan Haider.

My sincere thanks go to my incredible beta readers—Christof Mees, Apurva Desai, Ted Shelton, Jim Beverley, Penny Pearl, and Mindy O'Toole—whose valuable feedback elevated this work.

My deepest thanks go to the exceptional team at Manuscripts, LLC, who transformed this vision into a reality with their tireless efforts. I am particularly indebted to Chrissy Wolfe, revisions editor, and George Thorne, marketing lead, for their unparalleled customer focus and steadfast support. I also extend my gratitude to Gjorgji Pejkovski and Nikola Tikoski for their remarkable talent in crafting the captivating cover and illustrations.

This book would not exist without the remarkable community who preordered and promoted it enthusiastically. I am truly humbled and thankful to every one of you for your incredible support: George Alifragis, Richard Allaway, Theresa Amatore, Samir Balahouane, Rameshwar Balanagu, Tim Barnes, Lauren Berman, James Beverley, Jerry Briggs, Trinadh Bylipudi, Jennie Byrne, Jaime Cardenas, Larry Cassou, Seth Catalli, Wendy Chin, Murthy Chintalapati, Cheryl Chiovetta, Christy Clark, Jesse Craycraft, Scott Davis, Apurva Desai, Chaitany Desai, Priyanka Dobriyal, Deepak Dube, Huw Evans, Natalie Fay, Jane Feng, Sreenivasa Rao Ganapa, Vijaya Kumar Garimella, Jim Goldfinger, Madhu Goundla, Preethi Goundla, Govindan Anumarla, Linda Graebner, Claire Grosjean, Lance Grow, Farhan Haider,

Scott Herren, Kirk Holmes, Clayton Hull, Beth Hunt, Matthew Ikle, Beverly Ivens-Telepan, Princy Jain, Kiran Kanakadandila, Nikhila Kanakadandila, Jan Kang, Santosh Karre, Manoja Konduri, Lindsay Kulkin, Pranay Kumar, Herve Le Jouan, Erin Leixner, Oscar Luna, Gil Luria, Julie Mak, Dana Mallozzi, Zach Maritz, John McNiff, Rao Meda, Christof Mees, Kesava Merugumala, Talila Millman, Claus Moldt, Ajanta Nallamilli, Michelle Nasir, Jeanann Nichols, Lynn Olson, Oisin O'Reilly, Mindy O'Toole, Gregory Parker, Amanda Parness, Udaya Bhaskar Pasam, Sona Patadia-Rao, Krish Patel, Saagar Patel, Penny Pearl, Jerry Phillips, Ray Pineda, Anita Ramachandran, Mahendra Ramachandran, Nageswara Rao, Jonathan Rioux, Jay Roberts, Ken Robinson, Martha Sager, Federico Salvitti, Steven Santamaria, Catherine Schober, Sudan Sethuramalingam, Ted Shelton, Jesse Singh, Bill Siu, Tracy Smith, Divakar Somanchi, Kristine Stevenson, Sivaram Sunkara, John Sviokla, Joanne Tan, Mark Templeton, Jeff Tripaldi, Anurag Varshney, Minette Viljoen, Carola Volkmann, Matt Weinreich, Nageswara Rao Vysyaraju, Mariette Wharton, and Sondra Wudunn.

Your belief, guidance, and encouragement have been the driving force behind this accomplishment. Thank you all.

References

INTRODUCTION

1. World Economic Forum, "Recession and Automation Changes Our Future of Work, but There are Jobs Coming, Report Says," October 20, 2020, https://www.weforum.org/press/2020/10/recession-and-automation-changes-our-future-of-work-but-there-are-jobs-coming-report-says-52c5162fce/.
2. Mark Cuban, "Watch CNBC's Full Interview with Billionaire Investor Mark Cuban," interview by Morgan Brennan, *CNBC*, March 4, 2024, 12:06, https://www.cnbc.com/video/2024/03/04/watch-cnbcs-full-interview-with-billionaire-investor-mark-cuban.html.

CHAPTER 1

1. Paul A. Freiberger and Michael R. Swaine, "ENIAC," *Technology* (blog), *Encyclopedia Britannica*, May 3, 2024, https://www.britannica.com/technology/ENIAC.
2. John McCarthy et al., "Artificial Intelligence Coined at Dartmouth," *About* (blog), Dartmouth College, accessed July 7, 2023, https://home.dartmouth.edu/about/artificial-intelligence-ai-coined-dartmouth.
3. Mark Baard, "AI Founder Blasts Modern Research," *Science* (blog), *Wired*, May 13, 2003, https://www.wired.com/2003/05/ai-founder-blasts-modern-research/.

4   Louie Andre, "The Internet by Data in 2020," *Research* (blog), *Finances Online*, June 15, 2021, https://financesonline.com/how-much-data-is-created-every-day/#1.

5   CHESScom, "Kasparov vs. Deep Blue | The Match That Changed History," *Articles* (blog), Chess, Oct 12, 2018, https://www.chess.com/article/view/deep-blue-kasparov-chess.

6   History Editors, "Deep Blue Defeats Garry Kasparov in Chess Match," *This Day in History* (blog), *History*, accessed July 7, 2023, https://www.history.com/this-day-in-history/deep-blue-defeats-garry-kasparov-in-chess-match.

7   Murray Campbell, A. Joseph Hoane Jr, and Feng-Hsiung Hsu, "Deep Blue," *Artificial Intelligence* 134, no. 1–2 (January 2002): 57–83, https://doi.org/10.1016/s0004-3702(01)00129-1.

8   John Tromp and Gunnar Farnebäck, "Combinatorics of Go," (Conference Paper, 5th International Conference, Turin, Italy May 29–31, 2006), https://doi.org/10.1007/978-3-540-75538-8_8.

9   Demis Hassabis, "AlphaGo: Using Machine Learning to Master the Ancient Game of Go," *The Keyword* (blog), Google, January 27, 2016, https://blog.google/technology/ai/alphago-machine-learning-game-go/.

10 Google DeepMind, "AlphaGo," *Technology* (blog), *Google DeepMind*, May 14, 2024, https://deepmind.google/technologies/alphago/.

11 Demis Hassabis, "What We Learned in Seoul with AlphaGo," *The Keyword* (blog), Google, March 16, 2016, https://blog.google/technology/ai/what-we-learned-in-seoul-with-alphago/.

CHAPTER 2

1   PwC, *Sizing the Prize: What's the Real Value of AI for Your Business and How Can You Capitalise?* (New York: PwC, 2018), https://www.pwc.com/gx/en/news-room/docs/report-pwc-ai-analysis-sizing-the-prize.pdf.

2   George Hopkin, "AI Spending Growth Predicted to Continue for Five Years," *Data & Analytics* (blog), *AI Magazine*, September 14, 2022, https://aimagazine.com/articles/ai-spending-growth-predicted-to-continue-for-five-years.

3   Bernard Marr, "The Amazing Ways Toyota Is Using Artificial Intelligence, Big Data & Robots." *Innovation* (blog), *Forbes*, November 9, 2018, https://www.forbes.com/sites/bernardmarr/2018/11/09/the-amazing-ways-toyota-is-using-artificial-intelligence-big-data-robots/?sh=698959523863.

4   McKinsey Staff, "The State of AI in 2020," *QuantumBlack* (blog), McKinsey & Company, November 18, 2020, https://www.mckinsey.com/capabilities/quantumblack/our-insights/global-survey-the-state-of-ai-in-2020.

5   Tom Davenport and Nitin Mittal, *All-In on AI: How Smart Companies Win Big with Artificial Intelligence*, (Brighton, Massachusetts: Havard Business Review Press, 2023).

6   Siemens Staff, *MindSphere for Predictive Maintenance* (Munich, Germany: Siemens, 2019), 1–3, https://www.plm.automation.siemens.com/media/global/en/Siemens-MindSphere-for-Predictive-Maintenance-sb-76183-A4_tcm27-32272.pdf.

7   UPS, "UPS to Enhance ORION With Continuous Delivery Route Optimization," January 29, 2020, https://about.ups.com/us/en/newsroom/press-releases/innovation-driven/ups-to-enhance-orion-with-continuous-delivery-route-optimization.html.

8   Swati Kirti, "How Walmart Plans to Use AI to Reduce Waste," *Global Tech* (blog), *Walmart*, April 30, 2024, https://tech.walmart.com/content/walmart-global-tech/en_us/blog/post/how-walmart-plans-to-use-ai-to-reduce-waste.html.

9   Deutsche Telekom, "Artificial Intelligence at Deutsche Telekom," *Company* (blog), *Deutsche Telekom*, December 7, 2023, https://www.telekom.com/en/company/digital-responsibility/details/artificial-intelligence-at-deutsche-telekom-1055154.

10 Capgemini, "AI in Automotive Report," March 26, 2019, https://www.capgemini.com/news/press-releases/ai-in-automotive-report/.

11 Alec High, "2023 Hyperautomation Trends," *Home* (blog), *SalientProcess*, January 18, 2023, https://salientprocess.com/blog/2023-hyperautomation-trends/.

12   Amil Merchant and Ekin Dogus Cubuk, "Millions of New Materials Discovered with Deep Learning," *Research* (blog), *Google DeepMind*, November 29, 2023, https://deepmind.google/discover/blog/millions-of-new-materials-discovered-with-deep-learning/.

13   Frédéric Jallat, "How AI Could Dramatically Improve Cancer Patients' Prognosis," *The Conversation* (blog), *The Conversation*, December 24, 2023, https://theconversation.com/how-ai-could-dramatically-improve-cancer-patients-prognosis-216713.

14   Andrew Senior, John Jumper, and Demis Hassabis, "AlphaFold: Using AI for Scientific Discovery," *Research* (blog), *Google DeepMind*, January 15, 2022. https://deepmind.google/discover/blog/alphafold-using-ai-for-scientific-discovery/.

15   The Waymo Team, "Cities, Freeways, Airports: How We've Built a Scalable Autonomous Driver," *Waypoint* (blog), Waymo, May 18, 2022, https://waymo.com/blog/2022/05/howwevebuiltascalableautonomousdriver/.

16   Shreyas Sharma, "Tesla: The Data Collection Revolution in Autonomous Driving," *CISS AL Big Data* (blog), October 23, 2023, https://medium.com/ciss-al-big-data/tesla-the-data-collection-revolution-in-autonomous-driving-03e069b8ccfc.

17   Adam Levy, "How Netflix's AI Saves It $1 Billion Every Year," *Fox Business* (blog), *Fox Business*, June 19, 2016, https://www.foxbusiness.com/markets/how-netflixs-ai-saves-it-1-billion-every-year.

18   Pay Space, "How Walmart Is Using Machine Learning AI to Boost Retail," *Science & Technology* (blog), *Pay Space Magazine*, May 25, 2023, https://payspacemagazine.com/articles/how-walmart-is-using-machine-learning-ai-to-boost-retail/.

19   Ibid.

20   AI Gems, "Amper—AI Music for Content Creators," *Music* (blog), *AI Gems*, accessed November 15, 2023, https://aigems.net/site/amper.

21 Deloitte, "The Future of Work," *Services* (blog), Deloitte, April 29, 2024, https://www.deloitte.com/global/en/services/consulting/collections/future-of-work.html.

22 Todd Spangler, "Netflix Adds Generative AI to Competitive Risk Factors in Annual Report," *Newsletters* (blog), *Variety*, January 26, 2024, https://variety.com/2024/digital/news/netflix-generative-ai-risk-factors-annual-report-1235889385/.

23 Hyperight, "Deep Brew: Transforming Starbucks into an AI & Data-Driven Company," *Data & Innovation* (blog), *Hyperight*, June 30, 2021, https://hyperight.com/deep-brew-transforming-starbucks-into-a-data-driven-company/.

24 John Ashley, "American Express Adopts Nvidia AI to Help Prevent Fraud and Foil Cybercrime," *Nvidia* (blog), *Nvidia*, October 5, 2020, https://blogs.nvidia.com/blog/american-express-nvidia-ai/.

25 Amol Patil, "The Good, the Bad, and the Awful of AI in Aerospace," *AI* (blog), *Aerospace Manufacturing and Design*, September 14, 2023, https://www.aerospacemanufacturinganddesign.com/article/the-good-the-bad-and-the-awful-of-ai-in-aerospace/.

26 Swati Nikumb, "Understanding the End-to-End UX of the Sephora Virtual Artist App—a UX Case Study," *UX Collective* (blog), May 30, 2020, https://uxdesign.cc/understanding-the-end-to-end-user-experience-of-the-sephora-virtual-artist-app-product-try-on-d8ae3f8d1fcf.

27 GE Vernova, "GE Vernova Unveils New AI-based Software to Advance Industrial Sustainability and Operations Goals Simultaneously," February 28, 2024, https://www.ge.com/news/press-releases/ge-vernova-unveils-new-ai-based-software-to-advance-industrial-sustainability-operations-goals.

28 Mike Kaput, "How Spotify Uses AI (and What You Can Learn from It)," *Marketing AI Institute* (blog), Marketing AI Institute, January 26, 2024, https://www.marketingaiinstitute.com/blog/spotify-artificial-intelligence.

29 Michael Kennedy, "Zara's Fashionable Future: How AI Drives Innovation in Online Retail," *Artificial Intelligence* (blog),

ProjectMetrics, July 16, 2023, https://projectmetrics.co.uk/title-zaras-fashionable-future-how-ai-drives-innovation-in-online-retail.

30 Divya Jain, "A Glimpse under the Hood of Adobe's AI and ML Innovations: Adobe Sensei ML Framework," *Adobe Tech Blog* (blog), November 1, 2019, https://blog.developer.adobe.com/a-glimpse-under-the-hood-of-adobes-ai-and-ml-innovations-54c8155801a8?gi=48fa34271462.

31 JPMorgan, "AI and Model Risk Governance," *Technology* (blog), *JPMorgan*, May 29, 2023, https://www.jpmorgan.com/technology/news/ai-and-model-risk-governance.

32 QuantMinds, "The Latest in LOXM and Why We Shouldn't Be Using Single Stock Algos," *Quant Finance* (blog), *Informa Connect*, May 16, 2018, https://informaconnect.com/the-latest-in-loxm-and-why-we-shouldnt-be-using-single-stock-algos/.

CHAPTER 3

1 Nvidia, "About CUDA," *NVIDIA Developer* (blog), *Nvidia*, accessed July 6, 2024, https://developer.nvidia.com/about-cuda.

2 TELUS International, "Data Labeling Fundamentals for Machine Learning," *AI Data* (blog), *TELUS International*, September 13, 2022, https://www.telusinternational.com/insights/ai-data/article/data-labeling-machine-learning.

3 Parasvil Patel, "Geoffrey Hinton on the Algorithm Powering Modern AI," *Radical Reads* (blog), *Radical Ventures*, August 31, 2023, https://radical.vc/geoffrey-hinton-on-the-algorithm-powering-modern-ai/.

4 Marian Croak and Jeff Dean, "A Decade in Deep Learning, and What's Next," *The Keyword* (blog), *Google*, November 18, 2021, https://blog.google/technology/ai/decade-deep-learning-and-whats-next/.

CHAPTER 4

1 Dan Milmo, "ChatGPT Reaches 100 Million Users Two Months after Launch," *News* (blog), *The Guardian*, February 3,

2023, https://www.theguardian.com/technology/2023/feb/02/chatgpt-100-million-users-open-ai-fastest-growing-app.

2. Michael Chui et al., *The Economic Potential of Generative AI: The Next Productivity Frontier* (New York: McKinsey & Company, 2023), https://www.mckinsey.com/capabilities/mckinsey-digital/our-insights/the-economic-potential-of-generative-ai-the-next-productivity-frontier.

3. Ashish Vaswani et al., "Attention Is All You Need," *arXiv* (blog), *Cornell University*, August 1, 2023, https://doi.org/10.48550/arXiv.1706.03762.

4. Maximilian Schreiner, "GPT-4 Architecture, Datasets, Costs and More Leaked," *AI Research* (blog), *The Decoder*, July 11, 2023, https://the-decoder.com/gpt-4-architecture-datasets-costs-and-more-leaked/.

5. Julian Horsey, "Which Claude 3 AI Model Is Best? All Three Compared and Tested," *Geeky Gadgets* (blog), *Geeky Gadgets*, March 5, 2024, https://www.geeky-gadgets.com/claude-3-ai-models-compared/.

6. Sundar Pichai and Demis Hassabis, "Our Next-Generation Model: Gemini 1.5," *The Keyword* (blog), *Google*, February 15, 2024, https://blog.google/technology/ai/google-gemini-next-generation-model-february-2024/.

7. Randstad, "More than Half of Workers Believe AI Will Future-Proof Their Careers, but Only 13% Have Been Offered Such Training Opportunities—Randstad Data Reveals," *News and Events* (blog), *Randstad*, September 05, 2023, https://www.randstad.com/press/2023/over-50-believe-ai-will-future-proof-their-careers-only-13-have-been-offered-ai-training/.

8. Giles Bruce, "Health System to Employees: Double-Check Your AI Work," *Newsletters* (blog), *Becker's Health IT*, September 11, 2023, https://www.beckershospitalreview.com/digital-health/health-system-to-employees-double-check-your-ai-work.html.

CHAPTER 5

1. David Miller, "100+ Project Management Statistics & Facts (Updated 2024)," *Home* (blog), *ProProfs*, April 1, 2024, https://

www.proprofsproject.com/blog/project-management-statistics/.

CHAPTER 6

1. Anne Trafton, "Artificial Intelligence Yields New Antibiotic," *MIT News* (blog), MIT, February 20, 2020, https://news.mit.edu/2020/artificial-intelligence-identifies-new-antibiotic-0220.
2. Ibid.
3. Anne Trafton, "Artificial Intelligence Yields New Antibiotic," *MIT News* (blog), MIT, February 20, 2020, https://news.mit.edu/2020/artificial-intelligence-identifies-new-antibiotic-0220.
4. Altay Ataman, *"Data Quality in AI: Challenges, Importance & Best Practices in '24," Data* (blog), *AIMultiple*, January 3, 2024, https://research.aimultiple.com/data-quality-ai/.
5. Adam Davis et al., "Virtual Production—A Validation Framework for Unreal Engine," *Netflix TechBlog* (blog), August 10, 2022, https://netflixtechblog.com/virtual-production-a-validation-framework-for-unreal-engine-aab780b2f8c8.
6. Betsy Reed, "Amazon Ditched AI Recruiting Tool That Favored Men for Technical Jobs," *News* (blog), *The Guardian*, October 11, 2018, https://www.theguardian.com/technology/2018/oct/10/amazon-hiring-ai-gender-bias-recruiting-engine.
7. Heidi Ledford, "Millions of Black People Affected by Racial Bias in Health-Care Algorithms," *Nature* 574 no. 7780 (October 2019): 608–9, https://doi.org/10.1038/d41586-019-03228-6.
8. Adam Zewe, "In Machine Learning, Synthetic Data Can Offer Real Performance Improvements," *MIT News* (blog), MIT, November 3, 2022, https://news.mit.edu/2022/synthetic-data-ai-improvements-1103.
9. Xu Guo and Yiqiang Chen, "Generative AI for Synthetic Data Generation: Methods, Challenges and the Future," *arXiv* (blog), *Cornell University*, March 7, 2024, https://arxiv.org/html/2403.04190v1.
10. Anne Trafton, "Artificial Intelligence Yields New Antibiotic," *MIT News* (blog), MIT, February 20, 2020, https://news.

mit.edu/2020/artificial-intelligence-identifies-new-antibiotic-0220.
11 Salesforce, "Unlock, Analyze, and Act on Your Data," *Resource Center* (blog), *Salesforce*, accessed April 15, 2024, https://www.salesforce.com/resources/guides/data-strategy-playbook/.
12 Salvador Rodriguez, "Facebook to Be Slapped with $5 Billion Fine for Privacy Lapses, Say Reports," *Tech* (blog), *CNBC*, July 15, 2019, https://www.cnbc.com/2019/07/12/ftc-fines-facebook-5-billion-for-privacy-lapses.html.
13 IBM, "What Is Data Governance?" *Think* (blog), *IBM*, accessed April 15, 2024, https://www.ibm.com/topics/data-governance.
14 Uber, "How Uber Achieves Operational Excellence in the Data Quality Experience," *Data/ML* (blog), *Uber*, August 5, 2021, https://www.uber.com/blog/operational-excellence-data-quality/.
15 BBC News, "Google Apologises for Photos App's Racist Blunder," *Tech* (blog), *BBC News*, July 1, 2015, https://www.bbc.com/news/technology-33347866.
16 Differential Privacy Team, "Learning with Privacy at Scale," *Machine Learning Research* (blog), *Apple*, December 2017, https://machinelearning.apple.com/research/learning-with-privacy-at-scale.
17 Connor Hoffman, "How Capable Is Tesla's Autopilot Driver-Assist System? We Put It to the Test," *News* (blog), *Car and Driver*, April 30, 2021, https://www.caranddriver.com/news/a35839385/tesla-autopilot-full-self-driving-autonomous-capabilities-tested-explained/.
18 TestingXperts, "5 Key Black Box Testing Principles for AI Systems," *Blog* (blog), *TestingXperts*, September 28, 2023, https://www.testingxperts.com/blog/black-box-testing.
19 Anthropic, "Anthropic's Responsible Scaling Policy," *News* (blog), *Anthropic*, September 19, 2023, https://www.anthropic.com/news/anthropics-responsible-scaling-policy?ref=blog.heim.xyz.
20 NIST, "U.S. Artificial Intelligence Safety Institute Consortium (AISIC)," *Information Technology* (blog), *NIST*, April 15, 2024,

https://www.nist.gov/aisi/ artificial-intelligence-safety-institute-consortium-aisic.

CHAPTER 7

1   Philip Marcelo, "FACT FOCUS: Fake Image of Pentagon Explosion Briefly Sends Jitters through Stock Market," *AP Fact Check* (blog), *AP News*, May 23, 2023, https://apnews.com/article/pentagon-explosion-misinformation-stock-market-ai-96f534c790872fde67012ee81b5ed6a4.
2   Rob Thubron, "REvil Ransomware Group Will Hand Over Kaseya Attack Decrypt Key for $70 Million," *News* (blog), *TechSpot*, July 6, 2021, https://www.techspot.com/news/90310-revil-ransomware-group-hand-over-kaseya-attack-decrypt.html.
3   IBM, *IBM Security X-Force Threat Intelligence Index 2024* (Armonk, New York: IBM, 2024), 1–65.
4   Josh Fruhlinger, "Equifax Data Breach FAQ: What Happened, Who Was Affected, What Was the Impact?" *Security* (blog), *CSO*, February 12, 2020, https://www.csoonline.com/article/567833/equifax-data-breach-faq-what-happened-who-was-affected-what-was-the-impact.html.
5   Alex Krizhevsky, Ilya Sutskever, and Geoffrey Hinton, "ImageNet Classification with Deep Convolutional Neural Networks," *Communications of the ACM* 60, no. 6 (May 2017): 84–90, https://doi.org/10.1145/3065386.
6   Amy Kraft, "Microsoft Shuts down AI Chatbot after It Turned into a Nazi," *X-Scitech* (blog), *CBS News*, March 25, 2016, https://www.cbsnews.com/news/microsoft-shuts-down-ai-chatbot-after-it-turned-into-racist-nazi/.
7   Dan Goodin, "Researchers Trick Tesla Autopilot into Steering into Oncoming Traffic," *Information Technology* (blog), *Ars Technica*, April 1, 2019, https://arstechnica.com/information-technology/2019/04/researchers-trick-tesla-autopilot-into-steering-into-oncoming-traffic/.
8   Chase Theodos et al., *Cybersecurity Considerations for Generative AI* (Washington, DC: Ingalls Information Security, 2023), 1–3.

9 Times of Israel, "Israeli Researchers Bypass Facial Recognition Using AI-Generated Makeup Patterns," *Homepage* (blog), *Times of Israel*, October 1, 2021, https://www.timesofisrael.com/israeli-researchers-bypass-facial-recognition-using-ai-generated-makeup-patterns/.
10 Ionut Ilascu,"NVIDIA Confirms Data Was Stolen in Recent Cyberattack," *News* (blog), *BleepingComputer*, March 5, 2022, https://www.bleepingcomputer.com/news/security/nvidia-confirms-data-was-stolen-in-recent-cyberattack/.
11 R.A. Becker, "Cyber Attack on German Steel Mill Leads to 'Massive' Real World Damage," *Nova* (blog), *PBS*, January 8, 2015, https://www.pbs.org/wgbh/nova/article/cyber-attack-german-steel-mill-leads-massive-real-world-damage/.
12 Guoming Zhang et al., "DolphinAttack: Inaudible Voice Commands," (New York, NY, USA: Association for Computing Machinery, 2017), 103–17, https://doi.org/10.1145/3133956.3134052.
13 Fred Lambert, "The Big Tesla Hack: A Hacker Gained Control Over the Entire Fleet, but Fortunately He's a Good Guy," *Tesla* (blog), *Electrek*, August 27, 2020, https://electrek.co/2020/08/27/tesla-hack-control-over-entire-fleet/.
14 Melissa Heikkilä, "Google DeepMind Has Launched a Watermarking Tool for AI-Generated Images," *Artificial Intelligence* (blog), *MIT Technology Review*, August 29, 2023, https://www.technologyreview.com/2023/08/29/1078620/google-deepmind-has-launched-a-watermarking-tool-for-ai-generated-images/.
15 Ryan Morrison, "Adobe and IBM among Latest to Sign AI Watermarking Code," *AI and Automaton* (blog), *Tech Monitor*, November 23, 2023, https://techmonitor.ai/technology/ai-and-automation/adobe-and-ibm-among-latest-to-sign-ai-watermarking-code.
16 Microsoft, "Microsoft SEAL," *Microsoft Research* (blog), *Microsoft*, January 4, 2023, https://www.microsoft.com/en-us/research/project/microsoft-seal/.
17 Ingrid Lunden, "Duality Nabs $30M for Its Privacy-focused Data Collaboration Tools, Built Using Homomorphic Encryption," *Startups* (blog), *TechCrunch*, October 5, 2021,

https://techcrunch.com/2021/10/05/duality-nabs-30m-for-its-privacy-focused-data-collaboration-tools-built-using-homomorphic-encryption/.

18 Betty Bryanton, "Introduction to Privacy Enhancing Cryptographic Techniques: Secure Multiparty Computation," *Data Science Centre* (blog), *Canada Revenue Agency*, March 15, 2024, https://www.statcan.gc.ca/en/data-science/network/multiparty-computation.

19 Pierre Tholoniat, "Have Your Data and Hide It Too: An Introduction to Differential Privacy," T*he Cloudflare Blog* (blog), *Cloudflare*, December 22, 2023, https://blog.cloudflare.com/have-your-data-and-hide-it-too-an-introduction-to-differential-privacy.

20 Ethereum.org, "What Are Zero-Knowledge Proofs?" *Zero-Knowledge Proofs* (blog), *Ethereum*, April 22, 2024, https://ethereum.org/en/zero-knowledge-proofs/.

CHAPTER 8

1 McKinsey & Company, "Why Do Most Transformations Fail? A Conversation with Harry Robinson," *Transformation* (blog), *McKinsey & Company*, July 10, 2019, https://www.mckinsey.com/capabilities/transformation/our-insights/why-do-most-transformations-fail-a-conversation-with-harry-robinson.

2 Jamie Elliot, "Introducing Lightful's AI Squad," *Artificial Intelligence* (blog), *Lightful*, August 7, 2023, https://lightful.com/blog/artificial-intelligence/Introducing%20Lightful%E2%80%99s%20AI%20Squad.

3 World Economic Forum, "Recession and Automation Changes Our Future of Work, but There are Jobs Coming, Report Says," October 20, 2020, https://www.weforum.org/press/2020/10/recession-and-automation-changes-our-future-of-work-but-there-are-jobs-coming-report-says-52c5162fce/.

4 Jmt5790, "Talent Wins Games, but Teamwork…Wins Championships," *Leadership* (blog), *PennState*, March 19, 2017, https://sites.psu.edu/leadership/2017/03/19/talent-wins-games-but-teamwork-wins-championships/.

5   Amazon Web Services, "Machine Learning University," *Machine Learning* (blog), *Amazon Web Services*, accessed February 15, 2024, https://aws.amazon.com/machine-learning/mlu/.
6   Adobe, "Adobe Digital Academy," *Corporate Responsibility* (blog), *Adobe*, accessed February 15, 2024, https://www.adobe.com/about-adobe/corporate-responsibility/creativity/digital-academy.html.
7   Deloitte, "Deloitte AI Academy™ Builds Tailored Generative AI Curriculum in Collaboration with Renowned Universities and Technology Institutions for Deloitte Professionals and Clients," August 24, 2023, https://www2.deloitte.com/us/en/pages/about-deloitte/articles/press-releases/deloitte-ai-academy-builds-tailored-generative-ai-curriculum-in-collaboration-with-renowned-universities-and-technology-institutions-for-deloitte-professionals-and-clients.html.
8   Srini Raghavan, "3 Ways Moveworks and Microsoft Teams Use AI to Improve Employee Productivity," *Microsoft 365* (blog), *Microsoft*, March 1, 2024, https://www.microsoft.com/en-us/microsoft-365/blog/2023/07/10/3-ways-moveworks-and-microsoft-teams-use-ai-to-improve-employee-productivity/.

CHAPTER 9

1   Tate Ryan-Mosley, "The New Lawsuit That Shows Facial Recognition Is Officially a Civil Rights Issue," *Artificial Intelligence* (blog), *MIT Technology Review*, April 14, 2021, https://www.technologyreview.com/2021/04/14/1022676/robert-williams-facial-recognition-lawsuit-aclu-detroit-police/.
2   Nan Russell, "Ethics and Trust at Work," *Career* (blog), *Psychology Today*, May 19, 2017, https://www.psychologytoday.com/us/blog/trust-the-new-workplace-currency/201705/ethics-and-trust-at-work.
3   Corey Ruth, "IEEE Introduces New Program for Free Access to AI Ethics and Governance Standards," *News* (blog), *IEEE Standards Association*, January 17, 2023, https://standards.ieee.org/news/get-program-ai-ethics/.

4   NIST, "AI Risk Management Framework," *Information Technology Laboratory* (blog), *NIST*, January 26, 2023, https://www.nist.gov/itl/ai-risk-management-framework.

5   Catalina Devandas-Aguilar, "Day of the African Child: Protecting All Children's Rights and Investing in Children is Paramount for Africa," *News* (blog), *United Nations*, accessed May 2, 2024, https://violenceagainstchildren.un.org/news/day-african-child-protecting-all-childrens-rights-and-investing-children-paramount-africa.

6   Susanne Hupfer et al., "Women in the Tech Industry: Gaining Ground, but Facing New Headwinds," *Articles* (blog), *Deloitte Insights*, December 1, 2021, https://www2.deloitte.com/us/en/insights/industry/technology/technology-media-and-telecom-predictions/2022/statistics-show-women-in-technology-are-facing-new-headwinds.html.

7   Jeffrey Dastin, "Insight—Amazon Scraps Secret AI Recruiting Tool That Showed Bias against Women," *World* (blog), *Reuters*, October 10, 2018, https://www.reuters.com/article/idUSKCN1MK0AG/.

8   Alex Najibi, "Racial Discrimination in Face Recognition Technology," *Science Policy and Social Justice* (blog), *Harvard University*, October 24, 2020, https://sitn.hms.harvard.edu/flash/2020/racial-discrimination-in-face-recognition-technology/.

9   Jeff Larson et al., "How We Analyzed the COMPAS Recidivism Algorithm," *Articles* (blog), *ProPublica*, May 23, 2016, https://www.propublica.org/article/how-we-analyzed-the-compas-recidivism-algorithm.

10 Ziad Obermeyer et al., "Dissecting Racial Bias in an Algorithm Used to Manage the Health of Populations," *Science* 366, no. 6464 (October 2019): 447–53, https://doi.org/10.1126/science.aax2342.

11 Spencer Feingold, "The European Union's Artificial Intelligence Act, Explained," *Europe* (blog), *European Business Review*, March 30, 2023, https://www.europeanbusinessreview.eu/page.asp?pid=6695.

12. IBM Developer Staff, "AI Fairness 360," *IBM Developer* (blog), *IBM*, November 14, 2018, https://www.ibm.com/opensource/open/projects/ai-fairness-360/.
13. Google AI, "AI Principles," *Responsibility* (blog), *Google AI*, 2023, https://ai.google/responsibility/principles/.
14. Mehrnoosh Sameki et al., "Model Interpretability," *Azure* (blog), *Microsoft Learn*, February 29, 2024, https://learn.microsoft.com/en-us/azure/machine-learning/how-to-machine-learning-interpretability?view=azureml-api-2.

CHAPTER 10

1. Frank Plaschke, Ishaan Seth, and Rob Whiteman, "Bots, Algorithms, and the Future of the Finance Function," *Strategy & Corporate Finance* (blog), *McKinsey & Company*, January 9, 2018, https://www.mckinsey.com/capabilities/strategy-and-corporate-finance/our-insights/bots-algorithms-and-the-future-of-the-finance-function.
2. Rossum, "Five Ways Artificial Intelligence Benefits Invoice Processing," *Ebooks* (blog), *Rossum*, accessed January 12, 2024, https://rossum.ai/five-ways-ai-benefits-invoice-automation/.
3. Lotta Lundaas, "Top 9 Use Cases for AI in AP Automation," *Blog* (blog), *Vic.ai*, April 17, 2024, https://www.vic.ai/resources/top-9-use-cases-for-ai-in-ap-automation.
4. Workiva, "How Hershey Simplifies SEC Reporting, SOX, and ESG," *Company* (blog), *Workiva*, accessed March 29, 2024, https://www.workiva.com/customers/how-hershey-simplifies-sec-reporting-sox-and-esg#case-study-content-container.
5. PepsiCo, *2023 Annual Report* (Purchase, NY: PepsiCo, Inc., 2024).
6. AIpracticalGuide.com, "AI in Insurance Is Changing the Industry," *AI Applications* (blog), *AI Practical Guide*, accessed March 30, 2024, https://aipracticalguide.com/ai-in-insurance/.
7. PayPal Editorial Staff, "The Power of Data: How PayPal Leverages Machine Learning to Tackle Fraud," *Fraud and Risk Management* (blog), *PayPal*, December 22, 2021, https://www.paypal.com/us/brc/article/paypal-machine-learning-stop-fraud.

8 Anand Trivedi, "Capital One: Transforming Traditional Banking to an AI-First Experience," *Digital Innovation and Transformation* (blog), *Harvard Business School*, November 26, 2022, https://d3.harvard.edu/platform-digit/submission/capital-one-transforming-traditional-banking-to-an-ai-first-experience/.

9 Mastercard, "Mastercard Supercharges Consumer Protection with Gen AI," *Newsroom* (blog), *Mastercard*, February 1, 2024, https://www.mastercard.com/news/press/2024/february/mastercard-supercharges-consumer-protection-with-gen-ai/.

10 BlackRock, "Investor's Guide to Artificial Intelligence," *Thematic Insight* (blog), *BlackRock Advisor Center*, November 9, 2023, https://www.blackrock.com/us/financial-professionals/insights/investor-guide-to-ai.

11 Michael Adams, "Betterment Review 2024," *Investing* (blog), *Forbes Advisor*, January 5, 2024, https://www.forbes.com/advisor/investing/robo-advisor-betterment-review/.

12 Thomas Davenport and Randy Bean, "The Pursuit of AI-Driven Wealth Management," *AI in Action* (blog), *MIT Sloan Management Review*, July 7, 2021, https://sloanreview.mit.edu/article/the-pursuit-of-ai-driven-wealth-management/.

13 Coupa, "Get Visibility and Control over All Your Spend," *Products* (blog), *Coupa*, accessed January 02, 2024, https://www.coupa.com/products/spend-analysis.

14 Janice Lee, "Future of Procurement with AI," *Enterprise Resources* (blog), *Amazon Business*, June 02, 2023, https://business.amazon.com/en/discover-more/blog/procurement-future-with-ai.

CHAPTER 11

1 Marriott International, "Meet RENAI by Renaissance: The Pilot Program for Renaissance Hotels' New AI-Powered Virtual Concierge Service," *News Center* (blog), *Marriott International*, December 6, 2023, https://news.marriott.com/news/2023/12/06/meet-renai-by-renaissance-the-pilot-program-for-renaissance-hotels-new-ai-powered-virtual-concierge-service.

2   Liza Colburn, "Five AI Personalization Tools to Watch," *Articles* (blog), *Persado*, February 16, 2024, https://www.persado.com/articles/ai-personalization-tools/.

3   The Coca-Cola Company, "Coca-Cola Invites Digital Artists to 'Create Real Magic' Using New AI Platform," *Media Center* (blog), *The Coca-Cola Company*, March 23, 2023, https://www.coca-colacompany.com/media-center/coca-cola-invites-digital-artists-to-create-real-magic-using-new-ai-platform.

4   Hyperight, "Deep Brew: Transforming Starbucks into an AI & Data-Driven Company," *Artificial Intelligence* (blog), *Hyperight*, June 30, 2021, https://hyperight.com/deep-brew-transforming-starbucks-into-a-data-driven-company/.

5   Swati Nikumb, "Understanding the End-to-End UX of the Sephora Virtual Artist App—A UX Case Study," *UX Collective* (blog), May 20, 2020, https://uxdesign.cc/understanding-the-end-to-end-user-experience-of-the-sephora-virtual-artist-app-product-try-on-d8ae3f8d1fcf.

6   Wes Davis, "Apple Is Already Using Its Chatbot for Internal Work," *Apple* (blog), *The Verge*, July 23, 2023, https://www.theverge.com/2023/7/23/23804825/apple-gpt-chatbot-apple-care-siri-chatgpt.

7   Kateryna Cherniak, "AI in Customer Service Statistics: 50+ Actionable Insights for Your Technology-Driven Business Strategy," *Blog* (blog), *Master of Code*, April 16, 2024, https://masterofcode.com/blog/ai-in-customer-service-statistics.

8   SambaNova Systems, "OTP Selects SambaNova to Build Europe's Fastest AI Supercomputer," *Press* (blog), *SambaNova Systems*, November 11, 2021, https://sambanova.ai/press/otp-bank-selects-sambanova-systems-to-build-europes-fastest-ai-supercomputer.

9   Druid, "OTP Bank Improves Customer Support with the Use of DRUID AI Virtual Assistant," *Case Studies* (blog), *Druid*, March 30, 2023, https://www.druidai.com/case-studies/conversational-ai-chatbot-banking-otp-bank.

CHAPTER 12

1. Persado, "What is Motivation AI?" *Resources* (blog), *Persado*, March 13, 2024, https://www.persado.com/resource-library/what-is-motivation-ai/.
2. Liza Colburn, "AI in Marketing: Benefits, Use Cases, and Examples," *Articles* (blog), *Persado*, July 6, 2023, https://www.persado.com/articles/ai-marketing/.
3. Divya Jain, "A Glimpse under the Hood of Adobe's AI and ML Innovations: Adobe Sensei ML Framework," *Adobe Tech* (blog), November 1, 2019, https://blog.developer.adobe.com/a-glimpse-under-the-hood-of-adobes-ai-and-ml-innovations-54c8155801a8?gi=48fa34271462.
4. Surbhi Nahata and Aksheeta Tyagi, "How Does NLP Elevate Your Customer Service," *Blog* (blog), *Sprinklr*, March 19, 2024, https://www.sprinklr.com/blog/nlp-in-customer-service/.
5. Sparsh Sadhu, "How Artificial Intelligence Is Changing Influencer Marketing," *The SocialPilot Blog* (blog), *SocialPilot*, March 11, 2024, https://www.socialpilot.co/blog/ai-influencer-marketing.
6. Lal Pekin, "Top 10 Social Media Analytics and Reporting Tools in 2024," *Social Media Analytics* (blog), *Sociality.io*, August 27, 2019, https://sociality.io/blog/social-media-analytics-tools/.
7. Mike Schroepfer, "Community Standards Report," *ML Applications* (blog), *Meta*, November 13, 2019, https://ai.meta.com/blog/community-standards-report/.
8. John Moulding, "YouTube's Machine Learning Is Getting Better at Flagging Extremist Content," *Newswire* (blog), *Videonet*, October 19, 2017, https://www.v-net.tv/2017/10/19/youtubes-machine-learning-is-getting-better-at-flagging-extremist-content/.
9. Ella Steen, Kathryn Yurechko, and Daniel Klug, "You Can (Not) Say What You Want: Using Algospeak to Contest and Evade Algorithmic Content Moderation on TikTok," *Social Media + Society* 9, no. 3 (August 2023), https://doi.org/10.1177/20563051231194586.
10. Scott Davenport, "Unleash the Power of AI for SEO Content with Moz's PARS Framework," *Blog* (blog), *Thrive Business*

Marketing, November 10, 2023, https://thrivesearch.com/unleash-the-power-of-ai-for-seo-content-with-mozs-pars-framework/.

11 Sprinklr, "Burnley Football Club Wins Competitive Advantage, Increases Social Media Engagement by 54%," *Customer Stories* (blog), *Sprinklr*, accessed March 2, 2024, https://www.sprinklr.com/stories/burnley-football-club/.

CHAPTER 13

1 Delta Bravo, "We'll Help You Drive Higher Productivity from a New Generation of Workers," *Delta Bravo AI* (blog), *Delta Bravo AI*, accessed April 7, 2024, https://deltabravo.ai/.

2 Accenture, "Accenture's Internal Operations Automation Journey," *Case Study* (blog), *Accenture*, accessed April 7, 2024, https://www.accenture.com/gb-en/case-studies/about/internal-operations-automation-journey.

3 Kevin Oliver, "How to Use AI to Optimize Queries and Automate Resource Management," *IBM Blog* (blog), *IBM*, June 19, 2020, https://www.ibm.com/blog/how-to-use-ai-to-optimize-queries-and-automate-resource-management.

4 Jorge Amar et al., "AI-Driven Operations Forecasting in Data-Light Environments," *Our Insights* (blog), *McKinsey & Company*, February 15, 2022, https://www.mckinsey.com/capabilities/operations/our-insights/ai-driven-operations-forecasting-in-data-light-environments.

5 Automation, "Rakuten Chooses Zebra Autonomous Mobile Robots to Improve Warehouse Operations," *Articles & News* (blog), *Automation*, March 24, 2022, https://www.automation.com/en-us/articles/march-2022/rakuten-zebra-autonomous-mobile-robots-warehouse.

6 Jenny Glasscock, "Echo Global Logistics Goes beyond AI to Drive Value," *Supply Chains* (blog), *FreightWaves*, October 27, 2023, https://www.freightwaves.com/news/echo-global-logistics-goes-beyond-ai-to-drive-value.

7 CNS Media, "Amazon Pioneers Artificial Intelligence Machine Learning for Packaging Waste Reductions," *All News* (blog), *Packaging Insights*, January 6, 2022, https://

www.packaginginsights.com/news/amazon-pioneers-artificial-intelligence-machine-learning-for-packaging-waste-reductions.html.

CHAPTER 14

1. Peter Tunney, "Hospitality Recruitment: How Is AI Changing the Way We Hire?" *Recruitment Tips* (blog), *Hosco*, September 14, 2020, https://employers.hosco.com/blog/hospitality-how-is-changing-the-way-we-hire.
2. Payslip, "Payslip Automate, Standardize and Streamline Payroll Management at Kirby Group," *Case Study* (blog), *Payslip*, accessed April 11, 2024, https://payslip.com/resources/case-studies/kirby-group.
3. Vertigo, "HireVue: A Face-Scanning Algorithm Decides Whether You Deserve the Job," *Leading with People Analytics* (blog), *Harvard University*, April 12, 2020, https://d3.harvard.edu/platform-peopleanalytics/submission/hirevue-a-face-scanning-algorithm-decides-whether-you-deserve-the-job/.
4. Riddhi Shah, "Pymetrics—Using Neuroscience & AI to Change the Age-Old Hiring Process," *Digital Innovation and Transformation* (blog), *Harvard University*, December 4, 2019, https://d3.harvard.edu/platform-digit/submission/pymetrics-using-neuroscience-ai-to-change-the-age-old-hiring-process/.
5. Lorelei Trisca, "How to Use AI in Employee Onboarding: 7 Key Use Cases to Drive Success," *HR Insights* (blog), *Zavvy*, March 13, 2024, https://www.zavvy.io/blog/ai-employee-onboarding.
6. Michelle Gouldsberry, "The Pivotal Role of AI in Performance Management," *Performance Management* (blog), *Betterworks*, September 20, 2023, https://www.betterworks.com/magazine/ai-performance-management/.
7. Andrii Bas, "Reflektive Review: Analysis of Performance Management," *Reflective Review* (blog), *Peoplelogic*, accessed April 12, 2024, https://peoplelogic.ai/blog/reflektive-review.
8. Training Industry, "SelfStudy Previews AI-Powered Platform for Delivering Personalized Content and Adaptive Learning Programs," April 26, 2018, https://trainingindustry.com/

press-release/content-development/selfstudy-previews-ai-powered-platform-for-delivering-personalized-content-and-adaptive-learning-programs/.

9 Brenna McCarthy, "Using Conversational AI to Support Employees' Psychological Needs," *Healthcare* (blog), *IBM*, November 1, 2021, https://www.ibm.com/blog/conversational-ai-employee-care/.

10 HireVue, "Unilever Finds Top Talent with HireVue Assessments," *Case Studies* (blog), *Hirevue*, accessed February 2, 2024, https://www.hirevue.com/case-studies/global-talent-acquisition-unilever-case-study.

CHAPTER 15

1 Retool, "State of AI," *A 2023 Report on AI* (blog), *Retool.com*, accessed February 3, 2024, https://retool.com/reports/state-of-ai-2023.

2 Kedasha Kerr, "Using GitHub Copilot in Your IDE: Tips, Tricks, and Best Practices," *Engineering* (blog), *GitHub*, March 25, 2024, https://github.blog/2024-03-25-how-to-use-github-copilot-in-your-ide-tips-tricks-and-best-practices/.

3 FindmyAItool, "Tabnine Coding Assistant," *Tabnine* (blog), *findmyaitool*, accessed April 12, 2024, https://findmyaitool.com/tool/tabnine.

4 Testim, "Accelerate Test Automation with the Power of AI," *Testim.io* (blog), *Testim.io*, accessed April 12, 2024, https://www.testim.io/ai/.

5 Michael Chui et al., *The Economic Potential of Generative AI: The Next Productivity Frontier* (New York: McKinsey & Company, 2023), https://www.mckinsey.com/capabilities/mckinsey-digital/our-insights/the-economic-potential-of-generative-ai-the-next-productivity-frontier.

6 Demis Hassabis, "AlphaFold Reveals the Structure of the Protein Universe," *Research* (blog), *Google DeepMind*, July 28, 2022, https://deepmind.google/discover/blog/alphafold-reveals-the-structure-of-the-protein-universe/.

7 Amil Merchant and Ekin Dogus Cubuk, "Millions of New Materials Discovered with Deep Learning," *Research* (blog),

Google DeepMind, November 29, 2023, https://deepmind.google/discover/blog/millions-of-new-materials-discovered-with-deep-learning/.

8 Michal Rosen-Zvi, "Finding New Uses for Drugs with Generative AI," *IBM Research* (blog), *IBM*, May 26, 2023, https://research.ibm.com/blog/generative-ai-new-drugs.

9 Procter & Gamble, "Leveraging Technology to Improve the Lives of P&G Consumers," *Blogs* (blog), *Procter & Gamble*, July 11, 2022. https://us.pg.com/blogs/executive-talks-innovation-vittorio-cretella/.

10 Mark Tyson, "Nvidia Uses GPU-Powered AI to Design Its Newest GPUs," *GPUs* (blog), *Tom's Hardware*, September 2, 2022. https://www.tomshardware.com/news/nvidia-gpu-powered-ai-improves-gpu-designs.

11 Autodesk, "Generative Design AI," *Solutions* (blog), *Autodesk*, accessed February 27, 2024, https://www.autodesk.com/solutions/generative-design-ai-software.

12 Ansys, "Introducing Ansys AI: A New Chapter in Engineering Simulation," *Webinars* (blog), *Ansys*, accessed April 13, 2024, https://www.ansys.com/webinars/introducing-ansys-ai-a-new-chapter-in-engineering-simulation.

13 Materialise, "Empowering the Choice for Sustainability," *Impact* (blog), *Materialise*, accessed April 13, 2024, https://www.materialise.com/en/about/impact.

14 Airbus, "Skywise," *Services* (blog), *Airbus Aircraft*, July 29, 2021, https://aircraft.airbus.com/en/services/enhance/skywise.

15 Qualitas Technologies, "EagleEye® Inspection System," *Product* (blog), *Qualitas Technologies*, November 21, 2023, https://qualitastech.com/eagle-eye-inspection-system/.

16 Splunk, ".conf23: Splunk Introduces New AI Offerings to Accelerate Detection, Investigation and Response across Security and Observability," July 18, 2023, https://www.splunk.com/en_us/newsroom/press-releases/2023/conf23-splunk-introduces-new-ai-offerings-to-accelerate-detection-investigation-and-response-across-security-and-observability.html.

17 Robert Barron, "Watson AIOps in Action: An Incident Timeline," *IBM Blog*, IBM, May 20, 2021, https://www.ibm.com/blog/watson-aiops-in-action-an-incident-timeline/.

18 AquSag Technologies, "Pioneering the Future: ServiceNow's Leadership in IT Service Management and Business Workflow Automation," *AquSag Technologies* (blog), December 29, 2023, https://medium.com/@aqusag/pioneering-the-future-servicenows-leadership-in-it-service-management-and-business-workflow-38c2d6ca8874.

19 Darktrace, "Darktrace Unveils New Cloud-Native Security Solution Using AI to Provide Real-Time Cyber Resilience for Cloud Environments," *News* (blog), *Darktrace*, October 26, 2023, https://darktrace.com/news/darktrace-unveils-new-cloud-native-security-solution.

20 GE Vernova, "5 Steps to Reaching Smart Predictive Maintenance," *Blog* (blog), *GE Vernova*, accessed April 13, 2024, https://www.ge.com/digital/blog/5-steps-reaching-smart-predictive-maintenance.

CHAPTER 16

1 University Medical Center Utrecht, "AI Speeds Up Identification Brain Tumor Type," October 11, 2023, www.sciencedaily.com/releases/2023/10/231011182111.htm.

2 Berkeley Lovelace Jr. et al., "Promising New AI Can Detect Early Signs of Lung Cancer That Doctors Can't See," *Health* (blog), *NBC News*, April 11, 2023, https://www.nbcnews.com/health/health-news/promising-new-ai-can-detect-early-signs-lung-cancer-doctors-cant-see-rcna75982.

3 Isobel Asher Hamilton, "Google's DeepMind Created an AI for Spotting Breast Cancer That Can Outperform Human Radiologists," *Science* (blog), *Business Insider*, January 2, 2020, https://www.businessinsider.in/science/news/googles-deepmind-created-an-ai-for-spotting-breast-cancer-that-can-outperform-human-radiologists/articleshow/73070327.cms.

4 Dezső Ribli et al., "Detecting and Classifying Lesions in Mammograms with Deep Learning," *Scientific Reports* 8, no. 4165 (2018): 1–7, https://doi.org/10.1038/s41598-018-22437-z.

5   IBM, "IBM Watson Health Introduces New Opportunities for Imaging AI Adoption," *News* (blog), *IBM Newsroom*, November 30, 2021, https://newsroom.ibm.com/2021-11-30-IBM-Watson-Health-Introduces-New-Opportunities-for-Imaging-AI-Adoption.

6   University of Oxford, "AI Tool Could Help Thousands Avoid Fatal Heart Attacks," *News & Events* (blog), *University of Oxford*, November 13, 2023, https://www.ox.ac.uk/news/2023-11-13-ai-tool-could-help-thousands-avoid-fatal-heart-attacks.

7   Muhammad Mateen Qureshi and Muhammad Kaleem, "EEG-Based Seizure Prediction with Machine Learning," *Signal, Image and Video Processing* 17, no. 4 (June 2023): 1543–54, https://doi.org/10.1007/s11760-022-02363-4.

8   Margot Savoy, "IDx-DR for Diabetic Retinopathy Screening," *American Family Physician* 101, no. 5 (March 2020): 307–8, https://www.aafp.org/pubs/afp/issues/2020/0301/p307.html.

9   John Jumper et al., "Highly Accurate Protein Structure Prediction with AlphaFold," *Nature* 596, no. 7873 (August 2021): 583–89, https://doi.org/10.1038/s41586-021-03819-2.

10 Atomwise, "Behind the AI: Synthetic Chemical Benchmarks for Testing What Structure-Based AI Models Are Learning," *AI Drug Discovery* (blog), *Atomwise*, June 27, 2022, https://blog.atomwise.com/synthetic-chemical-benchmarks-for-testing-what-structure-based-ai-models-are-learning.

11 Yuta Imai et al., "AI Analyses Neuron Changes to Detect Whether Drugs Are Effective for Neurodegenerative Disease Patients," *NU Research Information* (blog), *Nagoya University*, July 14, 2022, https://www.nagoya-u.ac.jp/researchinfo/result-en/2022/07/20220714-01.html.

12 Healx, "Healx Receives IND and Orphan Drug Designation for Fragile X Clinical Trial," *News & Opinion* (blog), *Healx*, May 26, 2023, https://healx.ai/ind-fragile-x-clinical-trial/.

13 Gwangju Institute of Science and Technology, "Deep Learning Model to Predict Adverse Drug-Drug Interactions," May 4, 2022, www.sciencedaily.com/releases/2022/05/220504092943.htm.

14 Oleksandra Furman, "How AI Wearable Technology in Healthcare Helps Serve Patients Better," *AI/ML* (blog), *Mind Studios*, June 6, 2024, https://themindstudios.com/blog/ai-and-wearable-technology-in-healthcare/.

15 Philips, "Patient Information Center iX (PIC iX)," *Patient Monitoring* (blog), *Philips*, accessed April 19, 2024, https://www.usa.philips.com/healthcare/product/HCNOCTN171/patient-information-center-ix-pic-ix.

16 Jo Cavallo, "How Watson for Oncology Is Advancing Personalized Patient Care," *Issues* (blog), *The ASCO Post*, June 25, 2017, https://ascopost.com/issues/june-25-2017/how-watson-for-oncology-is-advancing-personalized-patient-care/.

17 Tas Bindi, "Woebot Uses Conversational AI to Deliver Cognitive Behavioural Therapy," *Innovation* (blog), *ZDNET*, June 7, 2017, https://www.zdnet.com/article/woebot-uses-conversational-ai-to-deliver-cognitive-behavioural-therapy/.

18 Jon LaPook, "Mental Health Chatbots Powered by Artificial Intelligence Developed as a Therapy Support Tool," *60 Minutes* (blog), *CBS News*, April 7, 2024, https://www.cbsnews.com/news/mental-health-chatbots-powered-by-artificial-intelligence-providing-support-60-minutes-transcript/.

19 Qventus, "Simplify Healthcare Operations with the Industry's Long-Standing AI Leader," *Qventus* (blog), *Qventus*, accessed April 19, 2024, https://qventus.com/.

20 Dave Muoio, "GYANT Hauls in $13.6M Series A for AI Care Coordination Tool," *MobiHealthNews* (blog), *Mobihealthnews*, July 15, 2020, https://www.mobihealthnews.com/news/gyant-hauls-136m-series-ai-care-coordination-tool.

21 Catwell, "Meet Corti, the AI Helping Emergency Dispatchers," *Internet of Things* (blog), *Element14*, January 18, 2018, https://community.element14.com/technologies/internet-of-things/b/blog/posts/meet-corti-the-ai-helping-emergency-dispatchers.

22 Jackson Nurse Professionals, "The Transformative Impact of AI on Healthcare," *Blog* (blog), *Jackson Nurse Professionals*, April 8, 2024, https://www.jacksonnursing.com/blog/the-transformative-impact-of-ai-on-healthcare/.

23. NS Medical Devices, "How Babylon Health Is Using AI to Provide Online Healthcare Services," *News* (blog), *NS Medical Devices*, June 25, 2019, https://www.nsmedicaldevices.com/news/babylon-health-ai-health-services/.

24. Heather Landi, "Biofourmis Expands AI-Enabled Virtual Care to Monitor Complex Chronic Conditions," *Health Tech* (blog), *Fierce Healthcare*, February 14, 2022, https://www.fiercehealthcare.com/health-tech/biofourmis-expands-ai-enabled-virtual-care-complex-chronic-conditions.

25. Chih-Chien Hung et al., "Conception of a Smart Artificial Retina Based on a Dual-Mode Organic Sensing Inverter," *Advanced Science* 8, no. 16 (August 2021), https://doi.org/10.1002/advs.202100742.

26. Chris Larson, "Talkspace CEO: AI Can Improve Therapist Performance, Boost Quality," *Digital Health* (blog), *Behavioral Health Business*, February 24, 2023, https://bhbusiness.com/2023/02/24/talkspace-ceo-ai-can-improve-therapist-performance-boost-quality/.

27. Case Western Reserve University, "HoloAnatomy® Software Suite," *Why HoloAnatomy®* (blog), *Case Western Reserve University*, accessed April 22, 2024, https://case.edu/holoanatomy/.

28. Osmosis Team, "Osmosis News: Supporting Your Students with Digital Medical Education Resources," *Osmosis News* (blog), *Osmosis*, accessed April 22, 2024, https://www.osmosis.org/blog/2020/03/11/supporting-your-students-with-digital-medical-education-resources.

29. Businesswire, "First Accredited Continuing Medical Education Course Filmed in VR," *News* (blog), *Businesswire*, December 3, 2018, https://www.businesswire.com/news/home/20181203005240/en/First-Accredited-Continuing-Medical-Education-Course-Filmed-in-VR.

30. A.Team, "Optimizing Nurse Assignments through Innovative Prototyping," *Case Studies* (blog), *A.Team*, accessed April 22, 2024, https://www.a.team/master-case-studies-collection/hca-healthcare.

CHAPTER 17

1. Hamna Waheed et al., "Deep Learning Based Disease, Pest Pattern and Nutritional Deficiency Detection System for 'Zingiberaceae' Crop," *Agriculture* 12, no. 6 (May 2022): 742, https://doi.org/10.3390/agriculture12060742.
2. Taranis, "Crop Intelligence at Scale," *What We Do* (blog), *Taranis*, accessed April 28, 2024, https://www.taranis.com/.
3. Agremo, "How Agremo Analytics Works," *Learn* (blog), *Agremo*, accessed April 28, 2024, https://www.agremo.com/how-agremo-analytics-works/.
4. Julia Kalanik, "Blue River Technology: The Field of Machine Learning," *Digital Innovation and Transformation* (blog), *Harvard University*, April 19, 2021, https://d3.harvard.edu/platform-digit/submission/blue-river-technology-the-field-of-machine-learning/.
5. AgriWebb, "Livestock Business Management that Moves Your Whole Farm Forward," *Homepage* (blog), *AgriWebb*, accessed April 28, 2024, https://www.agriwebb.com/move-your-whole-farm-forward/.
6. Benson Hill, "Cloud Biology® Provides Greater Accuracy and Accessibility," *Homepage* (blog), *Benson Hill*, accessed April 28, 2024, https://bensonhill.com/.
7. Growthsetting, "How Amazon Leveraged AI for Personalized Shopping Experiences," *AI Predictive Analytics* (blog), *Growthsetting*, November 9, 2023, https://growthsetting.com/ai-predictive-analytics/amazon-ai-personalization/.
8. Jessica Brown, "Smart Mirrors Reflect a New Era of In-Store Retail." *Retail* (blog), *Insight.Tech*, August 16, 2023, https://www.insight.tech/retail/smart-mirrors-reflect-the-future-of-retail.
9. Susan Caminiti, "How Walmart Is Using AI to Make Shopping Better for Its Millions of Customers," *Technology* (blog), *CNBC*, March 27, 2023, https://www.cnbc.com/2023/03/27/how-walmart-is-using-ai-to-make-shopping-better.html.
10. Vlad Kovalskiy "AI-Powered Marketing: Revolutionizing Customer Segmentation and Personalization," *AI-Powered Marketing* (blog), *Bitrix24*, October 27, 2023, https://www.

bitrix24.com/articles/ai-powered-marketing-revolutionizing-customer-segmentation-and-personalization.php.

11 Team IA, "Real-Time Adaptability: Elevating Retail Strategies through Price Optimization," *Blogs* (blog), *Impact Analytics*, November 2, 2023, https://www.impactanalytics.co/blog/real-time-adaptability-elevating-retail-strategies-through-price-optimization.

12 Indiaai, "AI-Based Video Analytics Platform Driving Sales for a Footwear Retailer in India," *Case Studies* (blog), *Indiaai*, September 12, 2021, https://indiaai.gov.in/case-study/ai-based-video-analytics-platform-driving-sales-for-a-footwear-retailer-in-india.

13 Owais Ali, "How Is AI Being Used in Space Exploration?" *Editorial Feature* (blog), *Azoquantum*, December 4, 2023, https://www.azoquantum.com/Article.aspx?ArticleID=474.

14 NASA, "Gateway Deep Space Logistics," *Deep Space Logistics* (blog), *NASA*, March 21, 2024, https://www.nasa.gov/gateway-deep-space-logistics/.

15 Gage Taylor, "New AI Algorithms Streamline Data Processing for Space-Based Instruments," *NASA Earth Science and Technology Office* (blog), *NASA*, August 22, 2022, https://esto.nasa.gov/new-ai-algorithms-streamline-data-processing-for-space-based-instruments/.

16 V. Venkataramanan, Aashi Modi, and Kashish Mistry, "AI and Robots Impact on Space Exploration," *Advances in Astronautics Science and Technology* (February 2024), https://doi.org/10.1007/s42423-023-00147-7.

17 Businesswire, "Earth AI Makes the First Greenfield Mineral Deposit Discovery Using AI," *News* (blog), *Businesswire*, October 26, 2023, https://www.businesswire.com/news/home/20231026831502/en/Earth-AI-Makes-the-First-Greenfield-Mineral-Deposit-Discovery-Using-AI.

CHAPTER 18

1 Rajkumar Venkatesan and Seb Murray, "5 Stages of AI Maturity in Marketing: A Blueprint for the Marketing Revolution," *Thought Leadership* (blog), *Forbes India*, December 26, 2023,

    https://www.forbesindia.com/article/darden-school-of-business/5-stages-of-ai-maturity-in-marketing-a-blueprint-for-the-marketing-revolution/90515/1.
2 Ibid.
3 Thomas Davenport and Randy Bean, "Portrait of an AI Leader: Piyush Gupta of DBS Bank," *AI in Action* (blog), *MIT Sloan Management Review*, August 31, 2021, https://sloanreview.mit.edu/article/portrait-of-an-ai-leader-piyush-gupta-of-dbs-bank/.
4 Ibid.

CHAPTER 19

1 Ryan Gross, "How the Amazon Go Store's AI Works," *Towards Data Science* (blog), June 1, 2021, https://towardsdatascience.com/how-the-amazon-go-store-works-a-deep-dive-3fde9d9939e9.
2 Intel, "Industrial Robotic Arms: Changing How Work Gets Done," *Industrial Robotic Arm Solutions* (blog), *Intel*, accessed May 2, 2024, https://www.intel.com/content/www/us/en/robotics/robotic-arm.html.
3 Sophia Barron, "17 Customer Service Chatbot Examples (& How You Should Be Using Them)," *Service* (blog), *Hubspot*, September 20, 2023, https://blog.hubspot.com/service/customer-service-chatbots.
4 Salman Bahoo, et al., "Artificial Intelligence in Finance: A Comprehensive Review through Bibliometric and Content Analysis," *SN Business & Economics* 4, no. 23 (January 2024), https://doi.org/10.1007/s43546-023-00618-x.
5 MIT, "Five Companies Investing in Upskilling the Workforce," *Global Opportunity Forum* (blog), *MIT*, April 21, 2022, https://goi.mit.edu/2022/04/21/five-companies-investing-in-upskilling-the-workforce/.
6 Erik Brynjolfsson, Tom M. Mitchell, and Daniel Rock, "What Can Machines Learn and What Does It Mean for Occupations and the Economy?" *AEA Papers and Proceedings* 108 (May 2018): 43–47. https://doi.org/10.1257/pandp.20181019.

7   Luke Mason, "Artificial Intelligence Will Transform Everything w/ Martin Ford," *FUTURES Podcast*, released January 10, 2024, 1:01:14, https://futurespodcast.net/episodes/59-martinford.

8   Demis Hassabis, "AlphaFold Reveals the Structure of the Protein Universe," *Research* (blog), *Google DeepMind*, July 28, 2022, https://deepmind.google/discover/blog/alphafold-reveals-the-structure-of-the-protein-universe/.

9   Ukeje Chukwuemeriwo Goodness, "Decoding How Spotify Recommends Music to Users," *Technology Explained* (blog), *Make Use Of*, July 4, 2023, https://www.makeuseof.com/decoding-how-spotify-recommends-music-to-users/.

CHAPTER 20

1   Ali Sepas, et al., "Algorithms to Anonymize Structured Medical and Healthcare Data: A Systematic Review," *Frontiers in Bioinformatics* 2 (December 2022), https://doi.org/10.3389/fbinf.2022.984807.

2   Jim Mehta, "Personalized Marketing and Customer Segmentation: How to Create Targeted Campaigns," *Blog* (blog), *Abmatic*, November 18, 2023, https://abmatic.ai/blog/personalized-marketing-and-customer-segmentation-how-to-create-targeted-campaigns.

3   Rebecca Pacun, "The All-Inclusive Guide to AI-Driven Ecommerce Product Recommendations," *Blog* (blog), *Prefixbox*, June 5, 2023, https://www.prefixbox.com/blog/ecommerce-product-recommendations/.

4   Bob Phibbs, "Best Cross Selling and Upselling in Retail Tips," *Blog* (blog), *The Retail Doctor*, accessed May 8, 2024, https://www.retaildoc.com/blog/best-cross-selling-sales-tips-for-retailers.

5   Relog, "Revolutionizing Urban Logistics: How Drones and Autonomous Vehicles Are Changing the Delivery Landscape," *Logistics* (blog), *Relog*, accessed April 20, 2024, https://getrelog.com/en/blog/revolutionizing-urban-logistics-how-drones-and-autonomous-vehicles-are-changing-the-delivery-landscape.

6 Stephen Pratt, "How AI Can Solve Manufacturing's Waste Problem," *Emerging Technologies* (blog), *World Economic Forum*, April 22, 2021, https://www.weforum.org/agenda/2021/04/how-ai-can-cut-waste-in-manufacturing/.
7 Mahnoor Imran, "AI and the Gig Economy Transformation," *Blogs* (blog), *Atomcamp*, March 24, 2023, https://www.atomcamp.com/ai-gig-economy/.
8 Dave Andre, "aiOS by Hyperspace: Revolutionizing Community AI through Peer-to-Peer Networking," *AI News* (blog), *All About AI*, April 1, 2024, https://www.allaboutai.com/ai-news/aios-hyperspace-community-ai-peer-to-peer-networking/.
9 David De Cremer and Garry Kasparov, "AI Should Augment Human Intelligence, Not Replace It," *Business and Society* (blog), *Harvard Business Review*, March 18, 2021, https://hbr.org/2021/03/ai-should-augment-human-intelligence-not-replace-it.